C.H.A.O.S.S.

C.H.A.O.S.S.

An Essay and Glossary for Students and Practitioners of Global Environmental Governance

Richard E. Saunier
Richard A. Meganck

A.A. BALKEMA PUBLISHERS Leiden/London/New York/Philadelphia/Singapore

Colofon

Cover image by H. Yang (UIUC, HST, NASA). Chaotic and unorganized, the beautiful NGC 604 contains over 200 emerging stars.
Image accessed November 2, 2002 at http://antwrp.gsfc.nasa.gov/apod/astropix.html.

ISBN 90 5809 704 8 (hardback)
ISBN 90 5809 705 6 (paperback)

Published by: A.A. Balkema Publishers, a member of Taylor & Francis Group plc
www.balkema.nl and www.tandf.co.uk

To Kenton R. Miller: critic, colleague, mentor, friend.

Acknowledgements

What is written here is not new and the ideas have been debated, acknowledged, criticized, supported, damned and praised by a long list of colleagues and mentors, most of whom we still count as friends. Over the years, in a great variety of meetings, bus commutes, long flights, cheap bars and, increasingly, in expensive hotels, as well as on hill sides and mountain tops, the faces and names have pretty much stayed the same. In the beginning the conversations were of conservation and the push for wilderness legislation, then came the intense interest on "the environment" and debates on sustainable development – and now, on environmental governance. Throughout it all there was plenty of helpful critique and support and we would be remiss if we did not at least acknowledge the value of the comments and criticisms of the most important. The good in here belongs to them – the problems still belong to us. Thus, in no particular order, subject matter, level of importance or agreement, we owe a debt of thanks to following individuals: Ariel Lugo, Arthur Heyman, Axel Dourojeanni, Boris Utria, Brian Thomson, Chris McKay, David Munro, Dianne Wood, Gilberto Gallopin, James Nations, Joanne Martin-Brown, Joshua Dickinson, Juan Jose Castro-Chamberlain, Kenton Miller, Kirk Rodgers, Larry Hamilton, Luis Bojorquez, Manual Ramirez, Martin Goebel, Michael Freed, Noel Brown, Pablo Gonzalez, Peter Jacobs, the late Peter Thatcher, Ralph Daley, Richard Sandbrook, Robert Goodland, Silvio Olivieri, Stephen Bender, Ted Trzyna, Walter Vergara, William Possiel, Yadira Soto, Yolanda Kakabadse, and Zoila Giron.

Comments from several reviewers triggered significant changes in the text while others provided us with necessary encouragement. These include Amb. Jack Vaughn, Amb. Luigi R. Einaudi, Amb. Michael King, Amb. Vicente Sanchez, Andras Szöllosi-Nagy, Andrea Merla, Gus Speth, Alfred Duda, Bart Schultz, Joyeeta Gupta, Len Berry, Mantha Mehalis, Mohamed T. El-Ashry, Rick Schulberg, Robin Rosenberg, Todd Baldwin, Peter Stroo, and Tom Lovejoy.

We used a great many individual and institutional sources to build the glossary and list of acronyms. We appreciate their work and they are named on page 49-53. Of course, none of this would have been possible without the administrative and logistical support provided by Marita Ciudad Real.

We also wish to acknowledge Janjaap Blom, from A.A. Balkema Publishers for his assistance in guiding us through the publishing process and Peter Stroo, the graphic designer at UNESCO-IHE Institute for Water Education who undertook the task of formatting and layout for both the cover and the body of this text. His steadfast progress through this laborious process allowed us to actually realize our goal of publishing this book in 2004. To both of these gentlemen we owe a rather large debt of gratitude.

Finally, we want to thank our spouses Gayle Saunier and Janet O. Meganck for their support and love through thirty years of friendship between our families.

Foreword

With globalization, a nation's developmental/environmental problems are increasingly inter-linked with the economies and policies of other countries as well as of local communities. As such, governance in today's world is organized at different administrative levels and policies at all these levels need to be closely coordinated and linked; a development that is referred to as multi-level governance. Three tendencies are increasingly visible, a tendency to regulate everything at a global level to create harmonized conditions and political viability, a tendency to address policy problems at the community level through stakeholder participation in order to achieve legitimacy and effective implementation, and a tendency to create the conditions in which private and other non-governmental actors can more actively participate in the governance process. Although these tendencies can pull in opposite directions, the centrality of governance through bilateral, multilateral and global interstate negotiations remains obvious, and such negotiations are becoming increasingly dominant as time passes. The recent challenges to the UN system by the United States should not be construed as a weakening of the power of the UN, but a recognition that if the US accepts and ratifies a UN agreement, it can be held accountable within its own borders for non-implementation of the agreement by non-state actors.

Given this reality, it is vital that students of international relations and international law, diplomats and other negotiators and observers understand the conditions and realities within which global policymaking is undertaken. These conditions and realities are what Saunier and Meganck treat in their present book, *C.H.A.O.S.S.* (an acronym for "Can humans and other species survive?"). The word is well chosen, for to the uninitiated the global policymaking process appears chaotic and confusing. Many of the conferences that are a part of this process are attended by more than a thousand delegates, with multiple plenary sessions and often dozens of non-plenary sessions and informal side-meetings. How decisions are actually taken often remains a mystery for those present in the room, since many of the decisions are prepared and pre-negotiated with a few key parties long before the plenary debate actually starts. This is also because, although countries are officially represented by negotiators and diplomats, in many of the more technical negotiations, scientists are sent who may be extremely well qualified in their own field, but are very "green" when it comes to figuring out the intricacies of international negotiation. For many negotiators, they are often faced with a *fait accompli*. Only those more intimate with the global process know how influential the UN rules of procedure are in the process of global diplomacy and those who know how to use and abuse these rules effectively have, in effect, control over the process and its outcomes.

The realities of negotiations are often not taken into account neither in international relations theories, international law discussions, nor in negotiation theories. It is here, that the present book: *C.H.A.O.S.S.* – fills in a vital gap in the available knowledge and in making the whole process more accessible to the future negotiator and researcher of the international processes. It combines an abstract essay discussing the nature of the chaos in global governance with a very practical glossary that immediately makes many of the terms accessible and comprehensible to the outsider.

From an international relations perspective, most global developmental and environmental problems are closely inter-linked and are essentially "wicked" or "unstructured" problems that reflect issues where there is poor alignment between the costs of taking action and the benefits from the action, where the science is limited in its ability to actually address the issue because of the high level of complexity and uncertainty, and where there is little or no agreement on the underlying values that should determine how problems are to be addressed. Under such circumstances, all global processes can be viewed as social learning processes and there is an expectation that such processes will eventually lead to problem solving. For many of us in the social sciences, there is a "method in the madness." This is described in the present book as the emergence of procedures that flow out of the "edge of chaos" where we now find ourselves.

Against this background, the essay in this book has a fresh approach, as it examines the international process from the perspective of those who participate in the international governance process. The context, process, architecture and implementation of diplomatic encounters is described in brief but colorful detail. The book uses the metaphor of chaos to describe the international arena where hardly any consensus definitions exist, even for key terms such as sustainable development and environment, where the concept of ecosystem is not understood, where the term ecology is used as an ideology rather than as science that has useful insights and where a lack of political will is used as an excuse for not getting our homework done. The book then refers to additional tools to help deal with the critical challenges in global environmental governance. The book argues that the lack of political will reflects absence of finished homework and the fact that the consensus is probably at best only symbolic. It then develops a strategy to deal with this challenge.

The book makes pleasant and informative reading and in itself is a very useful and rewarding effort. If at all one needs to find fault in the book, then that is in its confidence that terms and problems have singular definitions. This may sound strange to the social scientist for whom problems are socially constructed, perceptions are reality, and "normal" (as opposed to post-normal or public interest) science does not have solutions for so-called "wicked" problems. On the other hand, too much relativism may also cloud one's perspective and a confident, eloquently argued portrayal of the global problems can help to clear ones own doubts as to how these problems are to be addressed and may help us prepare to participate more effectively in the international negotiating arena.

Joyeeta Gupta
December 2, 2003
Delft, The Netherlands

Contents

C.H.A.O.S.S.: An Essay

Introduction

"There is no human failure greater than to launch a profoundly important endeavour and then leave it half done."
Barbara Ward

The idea for this essay and its accompanying glossary occurred to us as the global community prepared for the World Summit on Sustainable Development (WSSD) – an arduous, two-year process fraught with misunderstanding and acrimony.[1] Reaction to the Summit outputs only strengthened our wish to offer something that would help reinvigorate a productive conversation on the problems suffered by Earth and its inhabitants. After all, at one time they were of sufficient moment for governments to pay many millions of dollars for their resolution. To our knowledge, few, if any, of them have gone away.

Opinions concerning the relevance and efficacy of the World Summit on Sustainable Development vary, of course. As with the previous world meetings of this nature, some participants have said that not only did nothing of value occur, the dialogue required for that to happen did not exist.[2] Referring to the meetings of the WSSD, for example, President Hugo Chavez of Venezuela summarized it as a "dialogue of the deaf," a statement often repeated by governments and NGOs from the South.[3] In addition to a perceived lack of dialogue, others cite the assumed obstinacy and far from satisfactory efforts by governments to meet the goals set at earlier summits of this kind, as well as a failure to set urgent goals for the future and provide for those already established.[4] On the other side, officials of UN agencies, and some governments from the North, publicize the decisions to implement *Agenda 21* and the number of viable

1 The World Summit on Sustainable Development (WSSD) took place from August 26 to September 4, 2002 in Johannesburg, South Africa.

2 For a critique of the 1992 UNCED see Adil Najan, "An Environmental Negotiation Strategy for the South," *International Environmental Affairs* 7.3 (1995): 249-87 and for the 2002 WSSD see Tom Bigg, "The World Summit on Sustainable Development: Was it worthwhile?" Accessed at http://www.iied.org/wwsd/publish.html on 15 October 2003.

3 The quote by President Chavez is from Robin Pomeroy, "Earth Summiteers Cast Doubt on Future World Meets," *Reuters News Publication Service 4* September 2002. A view from NGOs in the South can be found in Anju Sharma, Richard Mahapatra and Clifford Polycarp, "Dialogue of the Deaf," *Down to Earth* (Delhi: Centre for Science and Environment, 2002) 25-33.

4 As given in a press release from the World Resources Institute, "WRI Expresses disappointment over many WSSD Outcomes." Accessed at http://newsroom.wri.org on 6 October 2003.

partnerships established to further sustainable development as signs of success.[5] Still, in an acknowledgement that serious problems do exist, the UN has announced a moratorium on such large environmental meetings.[6] However, the North-South, East-West, and bottom-up debates concerning global environmental governance (GEG) are not confined to large summits by any means. Indeed, they are reproduced in thousands of other gatherings that the summits both precede and follow.

Little doubt exists that stubbornness and biased attention to data are scattered among the diplomats, scientists, and concerned citizens who work on the issues central to these meetings. More often than not, though, bickering and bias help expose the real reasons for the gamut that a GEG idea must run. These include but are not confined to the complex differences in economic and social policies and cultural values among the nearly two hundred sovereign nations that can become involved in the issues, the application of "soft" and "hard" power preferences by major states,[7] and the thousands of non-governmental organizations that now help to frame the GEG agenda.

Proving themselves helpful in many ways, NGOs also bring with them a long list of added concerns and demands for attention.[8] Likewise, corporations, newly converted either by conscience or by economics, add an array of ideas rivaling the products and services they market. A large and growing number of disenchanted, not to say disenfranchised, individuals and groups outside the gates also now claim a voice. Caused as much by differences in what all sides are hearing as it is a difference in their ultimate goals, some of these groups prefer disruption to dialogue to get their points across.

Make-up of state delegations can also create its share of disorder. Rarely are the economists and attorneys who frequently form part of a delegation fully prepared to discuss biology and chemistry. Their colleagues, the chemists and biologists, by and large know little of diplomacy; and, despite

5 UN Department of Economic and Social Affairs, Division of Sustainable Development, "Partnerships for Sustainable Development." Accessed at wttp://www.un.org/esa/sustdev/partnerships/partnerships.htm on 13 November 2003.

6 Geoffrey Lean, "U.N. Creates Watchdog Group in Lieu of Future Summits," *London Independent*, 8 September 2002.

7 Robert Kagan, "Power and Weakness," *Policy Review on Line*. Accessed 15 August 2002 at http://www.policyreview.org/JUN02/kagan.html; also Andrew F. Cooper, John English and Ramesh Thakur, *Enhancing Global Government: Towards a New Diplomacy* (Tokyo: UNU Press, 2002).

8 Elizabeth Corell and Michele M. Betsell, "A Comparative Look at NGO Influence in Intergovernmental Environmental Negotiations: Desertification and Climate Change," *Global Environmental Politics* 1.4 (2001): 86-107.

near heroic efforts, none of these groups can fully understand and recommend completely effective remedies for the variety and potency of the problems we face.

Further, though present early on in the debates on international environmental management, the themes of human rights and poverty have begun to take on new weight and cracks in the wall of those who want to stay fixed on "the environment" writ as natural resource conservation and pollution control are beginning to show. For some, "sustainable development" in its many forms seems to have usurped the concerns of much of the environmental movement. Even for those who may feel that inequity and poverty are not a part of the environmental agenda, these two realities alter what we can and cannot do.

Correctly, the search for pathways to accommodate these many differences is high on the agenda of the diplomats and scholars active in the themes of global environmental governance.[9] However, the problems of environmental governance at a very practical level can also lie with words that are misunderstood and misinterpreted as well as with those that are known but not heeded. Thus, in addition to the scholarly work meant to elucidate and treat the complexities of GEG, tools are required that can help orient and facilitate the productivity of those directly involved in negotiating international instruments and of those who prepare and report on the many congresses, conferences and other meetings along the way. More specifically, it seems necessary to crack open the closed and often confusing language that leads to unnecessary complications, and that can make for decisions and negotiations based on partial or mistaken information.

Jargon and acronyms are the standard (and necessary) language of global environmental governance. We offer you PIC-INC. If you had to say the phrase, "Intergovernmental Negotiating Committee for the Preparation of the Conference of Parties of the Rotterdam Convention for the Application of the Prior Informed Consent Procedure for Certain Hazardous Chemicals and Pesticides in International Trade" at any interval short of every few years, you would most certainly opt for PIC-INC, its acronym. Because of PIC-INC and countless other bits of GEG jargon, acronyms, and

---

9 Ronald B. Mitchell, "Of Course International Institutions Matter: But When and How?" *Proceedings of the 2001 Berlin Conference on the Human Dimensions of Global Environmental Change: Global Environmental Change and the Nation State*, Frank Birdman, Rainier Brohm and Klaus Dingwerth. (eds.) PIK Report No. 80 (Potsdam: Potsdam Institute for Climate Impact Research, 2002) 16-25; and Ronald B. Mitchell, "International Environmental Agreements Defined." Accessed 15 October 2003 at http://www.uoregon.edu/~rmitchel/IEA/overview/definitions/htm.

abbreviations, it appeared to us that GEG students and even long-term practitioners could use an easily accessible and sizeable glossary of these things. Such a glossary and a listing of acronyms make up the bulk of this primer.

GEG is a continuing process that occupies the time and energy of thousands of dedicated people during hundreds of meetings. Available minutes of these meetings, however, record discussions that seem repetitive, disjointed, and, many times, misdirected.[10] The turmoil that arises between the official and the unofficial, the governmental and nongovernmental, and the boardroom and the street, too often reflects a curious blend of obscure, but important, scientific fact and popular, but meaningless, science fiction. Because of weaknesses in how many interested parties handle the most basic of GEG terms, we have included a short essay on some of the more important of these that should help enrich the debate if not clarify the issues.

None of this is meant to suggest a failure of the GEG process or of the range of instruments successfully concluded because of it. We have watched that process closely over a fair number of years, and respect the dedication and energy brought to it by a great many individuals and institutions. Yet, how this unique and valuable culture communicates within itself often seems a significant contributing factor in the problems – real and imagined – so often cited by others.

Consequently, rather than "failed", we choose to describe the current state of the global governance process as "chaotic," a portrayal that few who are active in it today would deny. However, though it is possible to depict the present system with the dictionary definition of chaos as "turmoil and turbulence," we do not want to emphasize those aspects of it here. We prefer, instead, to see it in terms of the complexity sciences where slight initial differences, imprecision, positive feedback for something less than optimal and increasing complexity take a system to a "place" at the edge of chaos where it will either stagnate in the channel it has created, die altogether, skid into ceaseless chaos, or escape to something better.[11] It is this last alternative that makes the "edge of chaos" a potentially constructive place to be – and that, to a large degree, is our message.

---

10 Hillary French, "Reshaping Global Governance," *State of the World 2002-Progress Towards a Sustainable Society*, Christopher Flavin et al. (eds.) (London: Worldwatch Institute and W.W. Norton and Company, 2002) 174-98.

11 Edward N. Lorenz, *"The Essence of Chaos"* (Seattle: University of Washington Press, 1993) 161-179; and M. Mitchell Waldrop, *Complexity: The Emerging Science at the Edge of Order and Chaos* (New York: Touchstone, 1993) 198-240.

Complexity concepts, of course, are often misinterpreted and misused from the point of view of mathematics and the physical sciences and we are aware of the damage we may do in the eyes of these disciplines. Our intent, however, is only to make a "metaphorical application" of their ideas.[12] We do not mean, in any way, to explore what mathematical modeling may say about global environmental governance. Nevertheless, it does seem to us that the richness of the ideas coming out of the complexity sciences can usefully describe GEG and, as a result, lead to a fruitful rethinking of the governance process.

We also understand that some of what we have written here will be of little value to the many diplomats and scholars who dedicate themselves to global environmental governance. If they happen to find this book in their hands, we suggest that they skip over the next section. However, we do want them to look at the remainder of the essay since it represents at least one view from the field that they may or may not be missing. And, for better or for worse, we offer it as an effort to help ensure that the "profoundly important endeavor" of global environmental governance not be left half done.

12 Stephen Kellert, "When is the Economy Not like the Weather? The Problem of Extending Chaos Theory to the Social Sciences," *Chaos and Society*, A. Albert (ed.) (Amsterdam: IOS press, 1995); and Robert Axelrod, *The Complexity of Cooperation* (Princeton: Princeton University Press, 1997) 3-9.

Context

"The art of international diplomacy is to convert the obnoxious into the incomprehensible"
Widely held principle of GEG

Global Governance

First a few words concerning a phrase you have already read many times in this document: "global environmental governance." Each of the three words is important. The concept of "governance" requires discussion because of the subtle but significant differences between it and the concept of "government."[13] Though they cover much of the same ground, governance is more than government. For example, according to the UNCHS, the differences are between "a single authority and shared purposes and responsibilities."[14] Further, governance includes all of the ways that individuals and institutions plan and manage their common affairs and consists of "formal institutions, informal arrangements," and what citizens know and do.[15] However, governance does not, as often believed, reduce the importance of a government since, as noted by the UNCHS, government still "holds the regulatory powers and the majority of fiscal responsibility;" and, its "normative and political legitimacy" helps to create and sustain the structures that encourage us to act together.

"Global," in the context of governance, likewise is different from "intergovernmental" and "international." While "intergovernmental" treats the official affairs between and among governments, "international," in addition to considering the relationships among governments, also includes the common contacts and dealings of their citizens. "Global" has a far broader meaning than either of the two previous ideas and embraces the official and unofficial governance activities of a long list of institutions including governments, businesses, non-governmental organizations,

---

13 Oran R. Young, "Global Governance: Drawing Insights from the Environmental Experience," *Global Environmental Accord: Strategies for Sustainability and Institutional Innovations*, O. R. Young (ed.) (Cambridge: MIT Press, 1997); Chapter Two of UNCHS, *Cities in a Globalizing World: Global Report on Human Settlements 2001*. (London: Earthscan Publications Ltd. 2002) 344 pp; and, Ronnie D. Lipschutz, "From Local Knowledge and Practice to Global Environmental Governance," *Approaches to Global Governance Theory*, Martin Hewson and Timothy J. Sinclair (eds). (Albany, NY: State University of New York Press, 1999) 259-83.

14 UNCHS p. 90-2, and the Commission on Global Governance at: http:/www.cgg.ch.

15 UNCHS p. 90-2.

universities, research centers, and foundations. Use of "global" acknowledges that a large number of institutions inside and outside of government and across national and institutional boundaries are responsible for much of the administration and management of our planet.[16] "Environmental" is a term of such importance that it has a section of its own and we encourage you to read the rest of the essay to see why.

As in all other issues of global governance, global environmental governance is founded on a few basic rules.[17] First, the fundamental premises outlined in the 1648 Westphalia Settlement concerning the absolute sovereignty and legal equality of nation states are at its center. Without these concepts, relationships between states in international forums would appear as little more than political parlor games and weight-lifting contests. Second, GEG follows international law – the compendium of binding agreements reached over years of interaction at various levels by which governments adjudicate differences and work toward positions of common interest and which become national law. In addition, GEG is guided by a growing number of principles (Appendix 1) and other non-binding instruments with the important caveat that, although agreed upon by consensus, these principles are not always accepted by all cultures in the same way.

A significant additional feature of global environmental governance is the role of civil society in decision-making accepted by governments at United Nations Conference on Environment and Development (UNCED) in Rio de Janeiro in 1992.[18] Principle 10 of the Rio Declaration states, in part, "Environmental issues are best handled with the participation of all

---

16 A number of works are available that seek to explain the ways and means of how this is done. Three among many are that of Patricia Birnie and Alan E. Boyle, *International Law & the Environment* (Oxford: Clarendon Press, 1992) 563 pp; the even larger and more wide-ranging *International Environmental Law and Policy* by David Hunter, James Salzman, and Durwood Zaelke, (New York: Foundation Press, 1998) 1567 pp; and that of Lamont Hempel, *Environmental Governance: the Global Challenge* (Washington D.C.: Island Press, 1996) 291 pp. Other frequently updated sources such as the *Earth Negotiations Bulletin* published by the International Institute for Sustainable Development and the Fridtjof Nansen Institute's *Yearbook of International Cooperation on Environment and Development* (London: Earthscan) for example, are excellent archival sources of facts and analysis. Of course, the treaties, and the records provided by their secretariats, fill in the necessary background and can be easily accessed through the United Nations Treaty Data Base at http://untreaty.un.org or the CIESIN Data Base at http://sedac.ciesin.columbia.edu.

17 For a comprehensive discussion of these "rules," see Chapters 1 and 2 of Birnie and Boyle (1992).

18 The documentation of UNCED in convenient form can be found in Michael Keating, *The Earth Summit's Agenda for Change: A plain language version of* Agenda 21 *and the other Rio Agreements* (Geneva: Centre for Our Common Future, 1993) 70 pp.

concerned citizens, at the relevant level." More importantly, governments enabled this process with the additional statement that, "States shall facilitate and encourage public awareness and participation by making information widely available. Effective access to judicial and administrative proceedings, including redress and remedy, shall be provided."[19]

Global environmental governance, which, in the run-up to the World Summit on Sustainable Development seems to have undergone a metamorphosis to become "global sustainable development governance," has at least three core components: process, architecture, and implementation. Ideally, GEG will reflect the interdependence of these components in that they should occur in a closely coordinated sequence. However, in the world of GEG, things may not work as they should and order may appear only after time, frequent false starts, and many frustrations.

The *process* of GEG can begin almost anywhere and often leads nowhere, as shown in the increasingly long list of unratified, totally abandoned or archaic instruments.[20] GEG *architecture* consists of the formal and informal, and public and private institutions that guide *implementation* – which often moves forward before it has a clearly defined destination or formal support.[21]

Process

The visible portion of the GEG process is made up of a large variety of assemblies, conferences, congresses and summits – so many in fact, that someone once defined sustainable development as the "mantra that launched a thousand meetings."Important examples of these meetings, of

---

19 The full text of the Rio Declaration can be found at http://www.unep.org/documents/default.asp?documentID=78&articleID=1163.

20 For example, the European Multilateral Agreement on Investments appears abandoned and the proposed Vienna Convention on the Law of Treaties between States and International Organizations or between International Organizations has seen little activity since 1986. Although accepted by UNGA RES 37/7 (1982), the largely NGO sponsored *World Charter for Nature* is looked upon as "pre-sustainable development" and insufficient to provide a complete ethical base for today.

21 This was particularly evident with early treaties but even with these problems, there were successes. The 1940 Western Hemisphere Convention on the conservation of natural resources for example, proved widely useful for the formation of numerous national parks and reserves in Latin America despite a lack of funding for a secretariat or any concerted official effort for its promotion. A discussion of the 1940 convention can be found in Kenton R. Miller, "International Cooperation Programs Which Relate to National Parks and Wildland Management," *Planning National Parks for Ecodevelopment: Methods and Cases from Latin America* (Washington D.C. Peace Corps Information Collection and Exchange, 1972) which can be accessed at: http://mng-unex1.marasconewton.com/peacecorps/documents/ROO73/r0073/r0073e/r0073e00.htm.

course, are the 1972 UN Conference on the Human Environment, the 1992 UN Conference on Environment and Development, and the 2002 World Summit on Sustainable Development. However, other meetings are now also considered as part of the GEG process; for example, the 1995 World Summit for Social Development, the 1996 Summit of the Americas on Sustainable Development, and the 2001 Doha Ministerial Conference of WTO. For every major meeting there are literally hundreds of other preparatory meetings sponsored by a galaxy of interests including those that offer the science that informs the negotiations.[22] These interests repeatedly break into sub-groups to deal with specific issues or aggregate into coalitions according to geography, economics, culture, ideology, and the question of the day. In formal meetings, a secretariat faithfully records the discussions and decisions for later review, debate, amendment, and action. A major part of the process is the effort to get it all down on paper in the proper language as an instrument that will enter into force when signed (by the executive branch of governments) and ratified (by the legislative branch of governments) by a sufficient number of states or other institutions.[23]

Meetings called on behalf of GEG can be large or small, formal or informal, inclusive or exclusive and may represent sectoral, geographical or cultural concerns. A speech, an article, or a background paper for any number of unofficial gatherings can initiate a new, or rekindle an old, GEG effort. The formal process established to negotiate an intergovernmental treaty or convention is a series of meetings of an intergovernmental negotiating committee (INC). However, rather than in official open meetings, negotiators often come to agreement first in hallways or during coffee breaks, cocktails, and "impromptu" late night working sessions where, despite the rhetoric, transparency can be an early casualty.[24] Similar meetings of NGOs (See Appendix 3) and, more recently, of the business

---

22 Victor Kremenyuk and Winfried Lang, "The Political, Diplomatic, and Legal Background," *International Environmental Negotiation*, Gunnar Sjöstedt (ed.) (London: Sage Publications 1993) 3-16.

23 The peculiarities of all of this can be confounding to newcomer and veteran alike. Neither should venture into a meeting without first reading Joyeeta Gupta's *"On behalf of My Delegation..." A Survival Guide for Developing Country Climate Negotiators*. (Climate Change Knowledge Network/Center for Sustainable Development in the Americas. Nd). Accessed on July 22, 2002 at http://www.unitar.org/cctrain/Survival%20Negotiators%(nAIP)www/index.htm. More general discussion of the whole procedure is given in Richard Elliot Benedick "Perspectives of a Negotiation Practitioner" in Sjostedt 219-43 and Birnie and Boyle 32-81.

24 This is not to denigrate the value of the debate that goes on in "unofficial" meetings where discussions can be much more straightforward than those that take place in plenary sessions.

community[25], mirror all of this in terms of complexity, cost, variety of interests, the amount and accuracy of information, and, secrecy.

Architecture

GEG architecture – the agreements and the institutions required to interpret and administer the instruments of governance is likewise, complex.[26] The organizations surrounding an agreement consist of a variety of formal and informal advisory committees (panels, subsidiary bodies, bureaus, etc.) as well as the conferences or meetings of the parties that negotiated and/or signed on to the treaty or protocol.[27] Secretariats, programs, and, commissions are part of this architecture. For example, the United Nations Environment Programme (UNEP) encourages and facilitates implementation of the mandates of the 1972 UN Conference on the Human Environment and the UN Commission on Sustainable Development (CSD) reviews progress on the approved agendas of the 1992 Earth Summit.[28] Most of the treaties and conventions that have entered into force now will also have a secretariat to track ratification, help insure implementation, and provide support to their administrative bodies.[29]

Agreements can be statements or declarations (sometimes defined as "soft law") and conventions, treaties, or protocols (sometimes defined as "hard law"). These different kinds of instruments bind the parties with different levels of authority.[30] A record of agreement for any of these instruments generally consists of an opening statement (*chapeau*), references to prior agreements, a set of principles, and statements regarding what the parties have agreed to do. Prescribed actions may follow as a part of the instrument itself, as an addendum, or as a separate text. Each institution and meeting has an identifying code for the documents that are produced (See Appendix 4 for help in deciphering the codes).

---

25 CCX, the Chicago Climate Exchange, a private consortium of 14 founding members recently signed a legally binding document (2003) agreeing to reduce greenhouse gases by 4% each in four years. See http://www.chicagoclimatetex.com.

26 See especially Chapter 1 and Chapter 2 of Birnie and Boyle for an excellent in-depth discussion of GEG architecture.

27 Commonly, Conferences of the Parties (COP) are administrative assemblies of parties to a convention or treaty while Meetings of the Parties (MOP) are administrative assemblies of parties to a protocol.

28 UNEP, *Compendium of Legislative Authority 1992-1997*. (Oxford: United Nations Environment Programme/Express Litho Service, 1993) 287 pp. and Keating.

29 Intergovernmental treaties and conventions are generally deposited with international organizations that will then dedicate an office to act as the secretariat.

30 Kenneth W. Abbott and Duncan Snidal, "Hard and Soft Law in International Governance." *International Organization*, 54.3 (2000) 421-56.

There are at least four varieties of conventions in GEG. Framework conventions, such as the UN Framework Convention on Climate Change and the Convention on Biological Diversity, require additional, separately negotiated protocols to reach the objectives laid out in the convention and a government may sign and even ratify a convention without signing or ratifying any of its protocols. Self-contained agreements such as UN Convention to Combat Desertification and the Convention on International Trade of Endangered Species do not necessarily need further protocols for implementation. Umbrella agreements, such as the Convention on Migratory Species of Wild Animals, allow other allied accords within the wider mandates of the convention. The fourth is an agreement open for signature to both governments and NGOs, as, for example, the Ramsar Convention on Wetlands.

Signing an instrument does not legally bind a nation to abide by its directives – only ratification can do that and only then if enough of its signatories ratify to allow entry into force.[31] However, signing does indicate that a state, or other institution, will do nothing to impede the work of the treaty secretariat or another state party toward meeting the objectives of the instrument that it signed.[32]

Implementation

As difficult and complex as the process and architecture of governance may be, they can appear easy compared to the implementation of what has been decided. Implementation of GEG frequently causes no end to anxiety, frustration, debate, accusations, and delays. All of this happens because implementation means financing the work of the agreement's plan of action; staffing of a secretariat or other institution(s) designed to insure that the agreement meets its objectives; and intense negotiations to decide just where the secretariat will be located and what the limits to its authority should be. National, and regional, prestige and additional income to a nation's economy are involved.

---

31 Ratification is seldom automatic. Indeed, if the treaty severely affects domestic economic interests, a legislative body may opt to not ratify even though the executive branch of the government signed the instrument (See Abram Chayes and Antonia H. Chayes, "Adjustment and Compliance Processes in International Regulatory Regimes," *Preserving the Global Environment: The Challenge of Shared Leadership*, Jessica Tuchman Mathews (ed.) (New York and London: W. W. Norton and Company, 1991) 280-308.

32 Fortunately, the *Vienna Convention on the Law of Treaties* signed at Vienna 23 May 1969 and entered into force 27 January 1980 governs all of this. Every negotiator should have a copy. It can be found at http://www.fletcher.tufts.edu/multi/texts/BH538.txt.

Debate on implementation also revolves around the issues of (a) funding for the programs mandated by a treaty and (b) sovereignty – with the developing world generally insisting that funding be "new and additional" and the bi- and multilateral development agencies insisting that certain conditions be met before the funding matters can be resolved.[33]

Monies for program and project implementation traditionally come from bilateral grants and low interest loans or multilateral loans that may be concessional in that they generally include lower interest rates and longer payback periods than those from commercial banks. More recently, funding and other assistance comes from NGOs and foundations, as foreign direct investments from corporations and other businesses or as some combination of public and private funds.

Whatever the case, funding is nowhere near what is required to fulfill current commitments to agreed upon targets for GEG. For example, the total annual official development assistance (ODA)[34] was between fifty and sixty billion dollars when the 180 governments attending the 1992 Earth Summit approved *Agenda 21*, a program that itself calls for 200 to 500 billion dollars annually. However, instead of increasing from 50 billion dollars to 500 billion dollars, or even 200 billion dollars, total ODA has dropped each year sense. It now sits at an average of 0.22% of GNP for the wealthier nations despite calls that it should rise to 1.0 % and agreement by these nations that it should be at 0.7% of GNP.[35]

Though much of GEG has little or no funding, some parts of it do.[36] The Global Environment Facility, for example, supports implementation of the Convention on Biodiversity, the Framework Convention on Climate Change, the Convention on Persistent Organic Pollutants, and the work of the Montreal Protocol Secretariat as well as work on international water

---

33 An important related issue is that of how agreements are to be enforced (See Ronald B. Mitchell, "International Environment," *Handbook of International Relations*. Thomas Risse, Beth Simmons, and Walter Carlsnaes (eds.) (London: Sage Publications, 2002); and Daniel W. Drezner, "Bargaining, Enforcement, and Multilateral Sanctions: When is Cooperation Counterproductive?" *International Organization*. 54.1 (2002) 73-102.

34 ODA is the official development assistance coming from the 22 members of the Organization for Economic Cooperation and Development (OECD).

35 In 2000, Denmark, at 1.06% of GNP had the highest percentage and the United States, at 0.10% had the lowest percentage of GNP dedicated to ODA. When presented in absolute amounts, the United States at $9.58 billion was second only to Japan at $13.06 billion in total ODA. An even more depressing view has recently been published by the Centre for Global Development and the Carnegie Endowment for International Peace, "Ranking the Rich," *Foreign Policy* (May June 2003) 56-66, which use other criteria to indicate support to the poor (trade, immigration and peacekeeping policies) in addition to aid and investment.

36 Hillary French, *Vanishing Borders* (London: Worldwatch Institute and W.W. Norton and Company, 2001).

and land degradation. Various trust funds support implementation of other individual agreements and ensure participation of delegates from Third World countries in many international meetings.

Fraying Edges

GEG, its process, architecture, and implementation are not static and in some thematic areas, both the substance and the rules that guide debate are rapidly evolving. Although governments generally adhere to the fundamental rules of how they are to interact, several recent actions seem to question the established codes of behavior.

- There is a trend toward greater polarization and hardening of positions between regions on certain issues – the rancorous debate over the Kyoto Protocol on global warming at COP-8 (October 2002) that has jeopardized the entire process is an example;
- Although science, rather than real or perceived power, was to drive the agenda of international negotiations, large and powerful countries and donor nations appear more aggressive in their attempts to influence the outcome of the GEG process for unilateral purposes;
- Movements like these force smaller countries to either align themselves with a larger country or vote as a geographic or interest-driven coalition to get a piece of the action. The Small Island Developing States (SIDS), for example, often votes as a group to garner the attention of the larger countries and to have a discernable influence on the outcome;
- In other instances, nations may "sell" (or "buy") votes in exchange for support, concessions or direct aid;
- Unilateral actions with regard to the Law of the Sea, the International Whaling Commission, and governments backing away from the Kyoto Protocol, International Criminal Court, Convention on Biological Diversity, Persistent Organic Pollutants, and the Basal Convention on the Movement of Hazardous Materials jeopardize core values of the United Nations;
- Disenfranchisement, perceived or real, of many of the world's citizens has caused opposition to global governance as the backbone of globalization. It is a significant uprising that will get stronger as it matures; and
- The different preferences (and capacities) by states to use "hard power" (military, economics) or "soft power" (negotiation) in the conduct of international affairs.

As a result, what may be emerging are the doctrines of "preemption" and "retroactivity" – even the use of threat and force that could significantly alter the GEG process. Some see a direct link between these and a so-called predominance doctrine or the willingness to go it alone on issues as disparate as war and intellectual property rights vis-à-vis the Convention on Biological Diversity.[37] In this scenario, power leads to unilateralist actions whether for vested interests (political, economic, etc.)or humanitarian or alturistic goals. A former Finnish Ambassador to the UN, noted that while the use of preemption as a strategy is historic, the context of employing it today has changed substantially.[38] Time will tell if these trends are merely opinions reflecting current events or if they will survive at the expense of what were once global values.

Success and Failure

Successes do occur despite, but not necessarily because of, the way institutions formulate and administer GEG. Almost paradoxically, success within the complexities of GEG seems to depend on several relatively simple attributes: personalities, relationships, trust, mutual respect, and a sympathetic understanding of others' interests. Rather than blocking coalitions, casting blame, and participating in ideological battles, joint problem solving is necessary. The process calls for high levels of expertise – technical as well as diplomatic – throughout the system. All participants must know and adhere to the decision-making procedures and a fixed and accepted set of rules. Conference officers should be stern, dedicated, unbiased, and knowledgeable of the history and substance of the issues. Also required are documents in the appropriate languages and a fully prepared secretariat, including translation and interpretation services as well as a cadre of dedicated note-takers. Besides these, the process should include the support, criticisms, ideas, and energy of civil society.[39]

Equally important, is that communication between the negotiators be efficient and with a minimum of error. Although, as is often the case, the negotiators may not agree on the definitions, as much as possible, all participants should seek to understand what the others are saying – even those populations who choose (or who are forced) to stay outside the

37 Max Boot, "The Big Enchilada: American Hegemony will be Expensive." *The International Herald Tribune*. 15 October 2002.

38 Max Jakobson, "Preemption: Shades of Roosevelt and Stalin." *The International Herald Tribune*. 17 October 2002.

39 James K. Sebenius, "The Law of the Sea Conference: Lessons for Negotiations to Control Global Warming" in Sjöstedt, 189-215.

gates. After all, what good is an instrument that contains all the right words if the parties that agreed to them understand them in mutually exclusive ways?

C.H.A.O.S.S.

"Squalid poverty lives side by side with over-abundance on our Earth. We have reached the Moon but we have not yet reached each other. Many species of our co-inhabitants on the globe . . . have forever disappeared. Many beautiful rivers have become sewers endangering the oceans. We must heed the omens. We must pass from words to deeds. We must pass from rights to obligations. We must pass from self-interest to mutual interest. We must pass from partial peace to total peace."

U Thant

Chaos

Though not the first expression of concern for the condition of our Earth and its inhabitants, these words of UN Secretary General U Thant over thirty years ago (1970) make an excellent place to start a discussion of C.H.A.O.S.S.. C.H.A.O.S.S. of course is an acronym. It corresponds to a question that requires an affirmative response from any effort to build an effective architecture of global environmental governance: "Can Humans and Other Species Survive?" Despite successes, the responses of the international community to the complexities that surround us have been messy and confused and the answers to the question have divided as often as they have united the peoples of our planet.

Arguments can be made that chaos is to be both expected and welcomed in something as complex as GEG, if for no other reason than that it creates the tension required for questioning and creativity to take place. Never-the-less, a case can also be made that this chaos, unattended, slows the process because of needless repetition, sidetracks the debate to unproductive bickering, raises the already excessive financial and human costs, and dampens the enthusiasm required to overcome the frequently tedious and thankless job of organizing a united attack on the problems we face today.

If GEG is looked at through the lens of the complexity sciences, it is the former rather than the later point of view that is of interest. Global environmental governance, like other social systems, is an open, non-linear, complex, and, therefore, unpredictable system that follows the trajectory:

Order → Complexity → Edge of Chaos → Chaos, disintegration, death, or emergence to a new order depending on the rules the system follows.[40]

Accordingly, a system begins its evolution from initial orderliness. However, as new information, projects, interests, misunderstandings, adjustments, constraints, and disturbances of any kind are added, the system becomes progressively more complex. It also becomes increasingly disorganized since, following the complexity sciences, small initial differences can later become major forces that push a system towards chaos or a variety of undesirable kinds of stability. Likewise, positive feedback, especially if for something less than optimal, will lead the system to increasing disorder and eventually to the "edge of chaos." It is here, at the edge of chaos, where an array of "choices" becomes available. "Attractors" – the points around which order is generated, influence these choices and there are three of them: "point," "cyclic," and "strange." If the system "falls" toward either of the first two, stagnation – even death – is the result. If it is the third, the consequence is chaos.[41]

Fortunately, though, there is a fourth "emergence" alternative, where creativity and innovation can take the system to something still complex but also newly ordered and, hopefully, in social systems, one having more value. Institutional examples of the first two attractors include all-important and immoveable organizational goals despite the change taking place around them. Cultures, core values and ideologies can be examples of strange attractors. With complex, open, non-linear systems, however, change or evolution of the system will always occur. Given the alternatives of death, eternal repetition, or endless chaos, it seems to many that the ideal policy would be one that influences, rather than attempts to stifle, this change.

Environment and Sustainable Development in GEG

GEG stands at the edge of chaos. Hints of this can be seen in the moves by individual states toward unilateralist actions, in the professionally belligerent, and, for our purposes, in the significant variation of estimates for the number of international environmental agreements (anywhere

---

40 Lorenz 228.

41 Lorenz 222-35.

from sixteen to nearly a thousand) given by GEG scholars and negotiators. These seemingly innocuous numbers indicate far more than innocent discrepancies because they depend on the informant's understanding of what the term *environment* may mean,[42] how it fits with his or her view of development,[43] and the relationship these concepts have with one's appreciation of what *sustainable development* may be.[44] Much of this confusion rests on the relationships between the terms of "conservation," "environmental protection," "environment and development" and "sustainable development" that have evolved over the last 50 years. This evolution, though normally positive, has also created many of the troubles that plague our search for successful models of global environmental cum sustainable development, governance.

Various GEG negotiators, and others, have long acknowledged the lack of consensus definitions for both "environment" and "sustainable development."[45] Lack of common, workable definitions of these terms, is one piece of the problem. Another is the growing distrust of sustainable development by some environmentalists and the accusations by non-environmentalist promoters of sustainable development that "representatives from the environmental ministries and departments dominate sustainable development." The following excerpt from the minutes of a recent meeting of a group of experts requested by the Open-ended Intergovernmental Group of Ministers illustrates the confusion:[46]

---

42 For example, Ronald B. Mitchell for the specific purpose of preparing a list of international environmental agreements, discusses this problem and defines agreements as being "environmental" if they ". . .seek, as a primary purpose, to manage or prevent human impacts on natural resources; plant and animal species (including in agriculture); the atmosphere; oceans; rivers; lakes; terrestrial habitats; and other elements of the natural world that provide ecosystem services." (*International Environmental Agreements Defined* accessed at http://www.uoregon.edu/~rmitchel/IEA/overview/definitions/htm on July 5, 2003). Ronnie D. Lipschutz, on the other hand, in his book *Global Environmental Politics From the Ground Up* (accessed at http://ic.ucsc.edu/~rlipsch/pol174/syllabus.html on July 15, 2003) prefers "habitat" as a more suitable approach to defining the global environment.

43 Herman E. Daly, *Beyond Growth: The Economics of Sustainable Development* (Boston: Beacon Press, 1996).

44 Richard E. Saunier, *Perceptions of Sustainability: A Framework for the 21st Century*.(Washington D.C.: Trends for a Common Future 6. CIDI/Organization of American States, 1999).

45 Becky J. Brown et al., "Global Sustainability: Toward Definition," *Environmental Management*, 11.6 (1987) 713-19; J. Pezzey, 1992, *Sustainable Development Concepts: An Economic Analysis*, World Bank Environment Paper No. 2, (Washington, D.C.: The World Bank, 1992); and Birnie and Boyle 2.

46 IISD, "Summary Report from the UNEP Expert Consultations on International Environmental Governance, 2nd Round Table, 29 May 2001" IISD *Linkages* 53.1 (2001).

> The Chair recalled that the Open-ended Intergovernmental Group of Ministers had asked for a clearer definition of international environmental governance and asked the experts to contribute to the debate on whether governance should refer to environmental issues or to sustainable development.
>
> The experts explored empowering the concept of sustainable development, to ensure the full integration of environmental policies with development. There was agreement that the problem is essentially political insofar as the key economic actors have not been engaged. This was taken as an indication that the sustainable development agenda lacks maturity. As a result, environmental actors are left to work only with the "converted" in trying to deal with the wider issues of sustainable development.
>
> This led to a critical discussion on the Commission on Sustainable Development (CSD), with one participant observing that it had been impossible to get down to real work because the only people in the conference room were ones with an environmental portfolio.
>
> There were differences of opinion over the question of the existence of a governance process that addresses sustainable development and that the only meaningful systems of governance address the environment. Some pointed out that the major environmental agreements also address sustainable development such as those designed to modify production and consumption patterns.
>
> Participants discussed the role UNEP could play in international environmental governance cum sustainable development governance. There was a consensus that, on the whole, the CSD adds little to the debate on sustainable development.

These comments reflect the diversity of understandings of environment and sustainable development extant in GEG – again, in large part, the result of fifty years of evolution. For example, a biologist at the time of the World Conservation Strategy[47] would have a different understanding of sustainable development from that of the forest manager in the 1950s or from an attorney at an environmental NGO who picked it up in the 1970s. All three of these interpretations are different from that of a diplomat who learned about it from the Brundtland Commission Report in 1987,[48] or a

---

47 IUCN, UNEP, and WWF, *World Conservation Strategy: Living Resource Conservation for Sustainable Development* (Gland: IUCN, 1980).

48 World Commission on Environment and Development, *Our Common Future* (Oxford: Oxford University Press, 1987).

business person who discovered the term in 1992 at the United Nations Conference on Environment and Development. These would be different from the view of almost anyone else who first learned about sustainable development during the Millennium Summit in 2000. All of these interpretations are possible, of course, but all of them are also incomplete. Taken together though, they are resources useful in our search for sustainability rather than something illegitimate or lacking in context.

GEG at the Edge of Chaos

Typically, the course followed in taking an idea through to a formal instrument of GEG begins with great enthusiasm and focus. Based on mutual concerns, states and/or other like-minded groups have common objectives that look down the line to the institutionalization of jointly made solutions. Negotiations commence but, almost immediately, positions begin to move away from one another. Multiple parties automatically mean multiple issues and unpredictable allies. The uncertainties of biased science and technology, whether real or imagined, begin to appear and, eventually, the realities of power differences become clear. Time frames evaporate and the cadres of early negotiators disappear – replaced by new people who perceive the potential outcomes as negative or competitive instead of common and convergent.[49] Negotiations continue and then, because of an immovable cut-off date, something less than originally hoped for is produced and signed.

Work towards global environmental governance requires a long institutional memory since the time counted from the recognition of a need for an agreement to its entry into force can easily exceed half of a career. By default, memory is held more by the community of observers (secretariat staffs and interested NGOs) than by the corps of official government negotiators for whom each new meeting seems a repetitive exercise for those who have been there, or a bizarre new learning experience for those who have just arrived. The numbing collection of concepts, jargon, and acronyms that the newcomer encounters will never quite be mastered before reassignment to different tasks in a different branch of the diplomatic corps.

Critics who say that official delegations to GEG events, including the meetings of the UN Commission on Sustainable Development, too often represent a country's environmental ministry or department have a point. However, the name "environment" hides important details – like the fact

49 Christophe Dupont, "The Rhine: A Study of Inland Water Negotiations." In Sjöstedt 135-48.

that the representatives can be located in, and reflect the view of, a variety of sectoral institutions that includes natural resources, agriculture, forestry, fisheries, health, education, culture, social development, housing, youth, labor, tourism, recreation, foreign affairs, finance, planning, science, cooperatives, energy, industry, and, even, public works. Now and then, they may even find themselves in the office of the chief of state where they can exercise genuine power. More often, they fill a more or less powerless office of their own.

Recognizing this diversity of content, delegations from the richer and larger nations have representatives from several different offices, including foreign affairs, who often disagree with their colleagues from the substantive units on the positions their government should take. Disagreements, however, rarely take place on technical grounds. Rather they are between the technical and political elements of the delegation.

Even if initially unified, the position of a delegation will vary over time. Governments, and a nation's policies, change with each election and, with each new government, a different set of negotiators arrives on the scene. During their initial meetings, at least, new delegates are unprepared for what they find despite what they may understand of their government's policies and positions on any given topic.

Generally, delegations contain both specialists and generalists. The predicament of the specialist is that the problem areas covered by GEG are often so broad that the specialist understands only a very small portion of the technical issues and almost nothing of diplomacy. The trouble with the generalist is that his or her understanding of a specific problem area is often nowhere close to what is necessary. Anything other than a solution based on an increasing number of vagaries required to reach consensus is impossible. Indeed, part and parcel of many of these agreements is the adjustment of once valid concepts and the invention of a language intentionally made unintelligible to everyone.[50] Agreement to use a term or concept that means different things to different people, however, can be a costly bit of trickery.[51]

---

50 Which caused Rosemary Righter to repeat an observation in a recent article in *The Times of London* (2003) that "The art of international diplomacy is to convert the obnoxious into the incomprehensible."

51 In diplomacy, "constructive ambiguity" in language is often used to get through a hard place so that negotiations can continue—maybe to a place where there is so much agreement that the point of ambiguity can be discarded altogether. Ideally, it has two characteristics: first, it is deliberate and all sides know and agree to what is going on and second, it reduces the differences to the lowest common denominator rather than to paper them over. It can be a dangerous tactic if the problem is not eventually resolved because the realities of the differences will always come back to cause even more difficulties.

The 1972 UN Conference on the Human Environment was the first international conference of its size to include a side gathering of a substantial number of civil society representatives. With each conference since then, civil society has moved closer to the official deliberations as well as to become better funded, larger in number, more varied in interests, better prepared and increasingly more aggressive in the promotion of its points of view.[52] As the influence of these groups grows, it adds both good and bad consequences in terms of complexity and results. Little doubt exists that the process achieves a better technical and, in many ways, political outcome because of the input and influence of civil society. However, the result often attempts to meet too many demands, appears disjointed, and almost certainly has a budget designed to meet negotiated objectives unacceptable to many governments that will claim they look more like a wish list than a realistic effort to find solutions that fit within national priorities and resources.

Not only are the non-governmental camp followers of GEG growing in number, they are also growing in variety and in strength sufficient enough to force new socio-economic topics like trade, debt relief, economic theory, the selection and assignment of subsidies, lifestyles, marketing, gender, war, democracy, and human rights onto the environmental agenda.[53] It is the flip side of the successes of the environmental movement that previously forced systems thinking, impact studies, health, and civil participation onto the agendas of development and governmental policy. Environmentalists may complain, but this is the logical extension of their early successes and, even, of the concept of environment itself.

Add to all of this the number of languages (including several different varieties of English) and the fuzziness created by the simultaneous interpretation of invented jargon. Add as well the heat created by cultural differences, ancient animosities, current competitions, and unfortunate histories. Add also the open and inadvertent skewing of information because of late night meetings, institutional loyalties, ephemeral alliances, and financial inequities. Include, then, the intricacies of working through national, institutional and delegate egos, as well as the delicacies of managing several simultaneous, cross-cultural flirtations, dalliances and

52 "Recent developments include the decisions to permit 'official non-government observers' at many sessions of intergovernmental organizations and meetings and, the development of formal mechanisms for civil society involvement in government decision-making and international meetings."

53 UNEP, *Global Environment Outlook 3*. (London: Earthscan Publications Ltd. 2002).

true loves (both institutional and personal), and there is little doubt that GEG stands at the edge of chaos.

We, the authors, of course, can do nothing about any of this. We have neither the authority nor the ability to do so. We do have, however, the opportunity to look at the language of GEG and we will do so in the next section.

Clarity and Chaos

"If language is not correct, then what is said is not what is meant.
If what is said is not what is meant, then what ought to be done remains undone.
If this remains undone, then morals and acts deteriorate and justice goes astray.
Hence, there must be no arbitrariness in what is said. This matters above everything."
Confucius

Chaos is Born

Mathematical models used in the development of the science of complexity show that chaos occurs in complex systems because of their sensitivity to initial conditions and, the models suggest, very small differences at the beginning of an operation will grow at an enormous rate. Likewise, as stated earlier, imprecision along the way, and positive feedback for something less than optimal will take an increasingly complex system to a "decision point" where it will either stabilize itself in death or monotony, stay in the furrow it has created to spend an eternity in chaos, or emerge to something still complex but once again ordered.

That such theories can describe social as well as physical systems is now widely accepted.[54] In the physical sciences, complexity and chaos can be explained because of the power of modern computers. For the social sciences, however, Confucius (see above) had it figured out long before the experts in non-linear programming.

Does Chaos Theory hold for global environmental governance? Our guess is, "yes." The differences and imprecision in how individuals and institutions in the environmental/sustainable development movement understand the words – not to mention the concepts – of "environment," "ecology," and "ecosystem" and the positive feedback surrounding the *politics* of "sustainable development" do seem to influence the course of GEG.

For many years, it has been a hobby of ours to ask students, teachers and technical and administrative staff of environmental management courses, government offices, and NGOs to define these four terms: environment, ecology, ecosystem, and sustainable development. No grades were involved but it is clear that few of the responses were coherent enough to

54 A. B. Cambel, *Applied Chaos Theory: A Paradigm for Complexity*. (San Diego: Academic Press, Inc. 1993).

serve as a guide to global environmental governance (See Appendix 4 for a collection of their responses). However, since even a professional could have a problem giving cold definitions to such requests, we need to look at what they give us in print, where they have sufficient time to reflect on what they say.

Environment

Table 1 presents the ways environmentalists often use the term "environment" when they talk or write to one another. In an attempt to classify the various uses, we have come up with four categories and several sub-categories of meanings. There could be many more. How the word is used is instructive, but the wide variation in how it is used (the "small initial differences" of Chaos Theory) is even more interesting. More importantly though, is that almost none of the uses really reflect the essential qualities of the scientific concept of environment: i.e. that the characteristics of an environment change in time and space and that these characteristics largely dictate the health, welfare and behavior of the object or system at its "center."[55]

Table 1: Contexts in which the word "environment" is defined.

I. Building Blocks	**II. Economic Uses**	**III. Spatial Uses**	**IV. Ethical/ Spiritual Uses**
Architectural	*Inputs*	*Ecosystems*	*Home*
Built environment	Natural Resources	Forest	Nature
Natural environment	System Services	Rangeland	Place
Geographical	*Outputs*	Planet	Planet
Terrestrial environment	Contamination	*Comprehensive*	Earth
Aquatic environment	Products	Watershed	*Spiritual*
Institutional	*Other*	Landscape	GAIA
Home environment	Occupational health		Wilderness
Work environment	Environmental Engineering		Culture
Social environment			Deep Ecology

55 Gilberto C. Gallopin, "The Abstract Concept of Environment," *Int. J. General Systems*, 7 (1981) 139-49.

Does this variety of uses cause problems for environmental governance? Again, our answer is "yes" and, this time, it is more than just a guess. For example, the confusion sown by being unacquainted with the concept of environment surfaces in the following discussion found in a UNDP report of a survey of environmental projects:[56]

> (p.6) As a first step, it was decided that UNDP would prepare a compendium of environmental projects, building on its own Official Development Assistance information base with contributions from forum members. The idea was to better understand the impact of aid to the environment sector...
>
> In 1996, UNDP published the *Compendium of Environmental Projects in Vietnam – 1985-1995* summarizing a decade of international assistance to the environmental sector...
>
> The team collating the compendium had to define an "environmental" project. This caused problems, as it is always difficult to define what activities constitute environment projects and programs.
>
> In the years immediately following [UNCED], aid agencies reported on their performance in this field using different criteria and by "repackaging" existing projects. The result was that it became impossible to gain an accurate picture of the true extent of environment investment...
>
> To get around this problem...UNDP adopted the definition that "an environmental project is one where the main objective of the project is either the preservation of the natural environment or supporting the sustainable management of natural resources."
>
> UNDP also noted that "even given this broad definition the problem of concisely classifying projects remains." It found that the best way to identify what is or is not an environment project was to illustrate the type of donor projects that did or did not fall into this category. Thus, donor projects identified as "environment" included:
>
> - All forestry projects;
> - All soil degradation and salinisation projects, minus irrigation activities;
> - Watershed and water pollution projects;
> - Environmental research and training, but not general agricultural research;

56 UNDP, *Compendium of Environmental Projects*. (New York: United Nations Development Program. 1996) accessed on 6 December 2003 at http://www.undp.org.vn/undp/docs/1996/zenvprojs/toc.htm.

- Integrated community and rural development projects, excluding commercially oriented agricultural and credit fund projects;
- Disaster planning and mitigation;
- Rehabilitation of sea dykes projects;
- Urban and regional development plans;
- Integrated pest management projects; and
- Energy efficiency projects (for example, wood stoves, solar or wind power).

The internal inconsistencies included in this list should be apparent to most readers. For example: do "all forestry projects" include high grading, clear-cutting, and the use of exotic species in plantations? Most urban and regional plans are terribly conflictive; integrated pest management is included in the list but agricultural research is not; salinisation projects [?] are in, but irrigation – the activity perhaps most responsible for salinisation – is out; and the debates surrounding the location of wind powered generators, waste treatment plants and sanitary landfills (pollution control) are classic in NIMBY wars as well as in the battles for environmental justice.

By no means, is such confusion isolated; nor is it new. In the early 1980's, the Committee of International Development Institutions on the Environment (CIDIE), which is made up of the environmental offices of UNEP, UNDP, the World Bank, IDB, ADB, AfDB, EIB, EBRD, OAS, and a few other part-time associates, was likewise concerned with isolating "environmental projects" from the rest of their agencies' portfolios. The group also needed a common basis for understanding the environmental management issues then confronting their agencies. Of the 35 issues suggested as being "environmental" by the nine CIDIE representatives (Table 2), two of them were mentioned seven times, four were mentioned three times, seven twice, and twenty-two of them were mentioned only once. No consensus was ever developed and the problem of what is and is not an environmental project remains unresolved. It is apparent that, in terms of its common usage, an "environmental project" is anything anyone wants to say it is.[57]

57 Birnie and Boyle have an interesting discussion on the topic of the many ways the word "environment" is used. Summarizing their discussion, they repeat the unfortunate words of Caldwell (1980): "...it [environment] is a term that everyone understands and no one is able to define." The words are unfortunate because the concept of environment is well understood in science and definitions that faithfully reflect that concept are available in any dictionary or encyclopedia.

Table 2: CIDIE Member Institutions' Interpretations of the Issues Requiring Treatment as an "Environmental Project."

Member Institutions	Interpretation
ADB	Reforestation, water supply, sanitation, solid waste disposal, urban renewal, land-use and capability assessment, watershed management, soil conservation, agro-forestry, industrial and urban pollution, regional planning.
EIB	Water supply, sewerage, wastewater treatment, oil tanker cleaning, industrial pollution, afforestation, pasture improvement and erosion control, soil saturation.
WB	Reforestation, soil conservation, rangelands and watershed management, water resources and slum upgrading, pollution control, natural resource development.
UNDP	Afforestation, energy conservation, pollution control, wildlife conservation, water supply and sanitation, flood warning, occupational safety and health, environmental legislation, environmental management surveys and studies.
IDB	Conservation or enhancement of the environment, flood control, land erosion, air, land and water pollution, transfer and utilization of science and technology.
AfDB	Sanitation, health, water, forestry, irrigation, agro-industry, mining, rural development, industry and energy.
ECE	Desertification and afforestation, wildlife conservation, renewable sources of energy, water management, village water supplies, soil conservation, fish stocks conservation.
OAS	Any development project is an environmental project.

Other contradictions are evident in how laymen and professionals alike understand the term "environment." Most will say, for example, that environment has to do with "natural resources," as in the extract of the UNDP report given above, but many others are fond of using the phrase "environment and natural resources" which seems to say that environment and natural resources are separate entities. There also is the widely accepted idea that "environment is everything" (See Appendix 5) which, if true, complicates things for those who prefer to talk about an "environmental sector," again as in the 1996 UNDP report.

Perhaps there are practical and operational reasons for many of the ways the term "environment" is used. Our conclusion, however, is that this word is the most used, yet least understood, term employed in environmental management as well as in the search for sustainable development and global governance. It has been misused for decades, of course, and too few writers use it in a way that sufficiently expresses what it has always meant: *Anything outside an object (or system) of interest that influences the health, welfare and behavior of that object (or system) of interest at a given moment in time.*[58]

Human environments, of course, are no different from other environments in terms of definition; they also are predominately local, always personal, continuously changing, and are a primary influence on the health, welfare and behavior of the individual or population at their "center." Therefore, any use of the term "human environment" that faithfully reflects the concept of environment will convey the idea that outside of naming a specific time, a specific place, and a specific owner, something called "the environment" does not exist.

Ecology

When asked to give a definition of "ecology" most people will get it more or less right and call it a "science." They will then immediately use the term to mean something entirely different.[59] An interesting example of the correct definition and incorrect use of the term appears in a publication of the Asian Development Bank in which "concepts of ecology" and "protection of ecology" show up in the same sentence after earlier defining the term only as the "...study of the interrelationships of organisms to their environment."[60]

Ecology is, indeed, a science – at least it is a science according to the *Ecological Society of America* (ESA), which, despite its name, is an association of professional ecologists from around the world. The full definition given by the ESA is "*Ecology is the scientific discipline that is concerned with the relationships between organisms and their past, present, and future environments.*"[61] Note the plural "environments." Note also that ecology is not a synonym for environment and that there is only one

58 Gilberto C. Gallopin, "Human Systems: Needs, Requirements, Environments and Quality of Life" In G.E. Lasker (ed.) *Applied Systems and Cybernetics. Vol. 1. The Quality of Life: Systems Approaches.* (Oxford: Pergamon Press, 1981); and Amos H. Hawley, *Human Ecology: A Theoretical Essay*.(Chicago: The University of Chicago Press, 1986).

59 Mojan Wali in the June 1995 *Bulletin of the Ecological Society of America* (Vol. 76: 106-11) lists 337 different uses of the terms "ecology" and "ecological" found in the published literature he reviewed.

60 Asian Development Bank, "Environmental Guidelines for the Development of Ports and Harbours." (Manila: ADB. 1986) 3.

61 Http://www.esa.org/aboutesa/

definition given. Used in any other way – as if "ecology" were a worldview, a synonym for environment or as an adjective to describe a problem, is "language that is not correct." To use it in any other way, without giving a definition of what one means by it, is confusing and impedes our capacity to "do what needs to be done."

Nevertheless, many of its users, though continuing to define it as a science, will say that the public long ago co-opted the word "ecology" and that one can now correctly use it to mean other things.[62] There is no doubt that definitions change because a word is misinterpreted and used erroneously so often that it eventually comes to mean something else. Alternatively, the definition can be modified because new information shows it to be in error as a descriptor for some phenomenon or object. Unfortunately, where ecology is concerned, co-optation is closer to the first than it is to the second – which brings up a question: Why should global environmental governance accept the careless use of words or, even more importantly, why should it accept the *ideological* use of words that, by definition, were designed to be free of ideology? If the word "ecology" is co-opted by an ideology, does the science itself also become ideological in the eyes of many? Observation tells us that the answer to that question is also a regrettable "yes."

Ecosystem

Even ecologists have trouble saying what an ecosystem is, yet the word is used by almost everyone else both as if they know for sure what it means and as if everybody who is listening has a perfect understanding of what is being said.[63] However, as with "environment" and "ecology," the popular view of "ecosystem" omits some very important parts.

For instance, according to the generally accepted *concept*, the borders of an ecosystem are not a given, are never precise, and, more importantly, are always defined by the person who locates the boundary around the ecosystem in question:[64] "[the] boundary is based on the observer's needs, the actors, and the context, rather than on any intrinsic property of the

---

62 Lynton Caldwell. *Between Two Worlds: Science, the Environmental Movement, and Policy Choice*. (Cambridge: Cambridge University Press. 1990) 239 pp.

63 John E. Fauth. "Working toward Operational Definitions in Ecology: Putting the System back into Ecosystem." *Bulletin of the Ecological Society of America*. 78.4 (1997): 295 and Alice E. Ingerson. "A Critical User's Guide to 'Ecosystem' and Related Concepts in Ecology." Institute for Cultural Landscape Studies, the Arnold Arboretum of Harvard University. Accessed June 10, 2002 at http://www.icls.harvard.edu/ecology/ecology.html.

64 John C. Maerz, "Ecosystem Management: A Summary of the Ecosystem Management Roundtable of 19 July 1993" *Bulletin of the Ecological Society of America* 75.2 (1994) 93-5.

system itself."[65] Other people, for other purposes or using other measurement instruments and scales, will put the line elsewhere.[66] Likewise, ecosystems, being non-linear and open, are neither static nor in equilibrium. They are always changing and this change can be rapid or slow, catastrophic or evolutionary, human controlled and influenced, or without any human intervention whatsoever.[67] These characteristics of ecosystems, of course, cause problems for such concepts as "ecosystem balance," "ecosystem health," and "ecosystem fragility" as well as for some interpretations of "ecosystem restoration."

For example, one of the myths surrounding the issue of sustainable development closely ties sustainability to an assumed "balance of nature" where disturbances initiated by humans lead to disharmony and development failure. However, science has discovered such balance, when it occurs, to be highly dependent on the scales of time and space.[68] Indeed, for all practical purposes, this kind of balance does not last and development, like environmental management, if it is to be sustainable, must either adapt to such change or fail – the reverse of what is often stated.

Further, though science *has* looked into the stability of ecosystems and, therefore, their instability under conditions of stress,[69] few studies have truly investigated their "fragility." Yet, there is a long list of ecosystems that have been labeled as fragile in the public press: from the Arctic to the tropics, from reefs and estuaries to alpine ponds, rainforests, grasslands, deserts, and incidentally, urban neighborhoods. Given its ubiquity in popular writing and its absence in scientific literature, however, one could suspect that the term *"fragile ecosystem"* is used more often to promote a

65 There is some debate on this issue but the broad outcome seems to be that of general systems theory which suggests that we isolate systems mentally for the purpose of study. See Victor Marín. "General System Theory and the Ecosystem Concept" *Bulletin of the Ecological Society of America* 78.1 (1997) 102-103 and Bernard Pavard and Julie Dugale. "An Introduction to Complexity in Social Science" p.17. Accessed July 15 2003 at http://www.irit.fr/cosi/training/Complexity-tutorial/htm.

66 Take, for example, the many differing definitions of "wetland" being debated for regulating land use. Closer to the subject of GEG are the discussions of Article 3.3 of the Kyoto Protocol and how one distinguishes between "forested," "reforested," and "afforested" areas.

67 James J. Kay, "A Non-equilibrium Thermodynamic Framework for Discussing Ecosystem Integrity" *Environmental Management*, 15.4 (1991) 483-95.

68 Daniel B. Botkin, *Discordant Harmony: A New Ecology for the Twenty-first Century.* (Oxford: Oxford University Press. 1990): 256 pp.

69 C. S. Hollings, "Resilience and Stability of Ecological Systems." *Annual Review of Ecology and Systematics* 4 (1973) 1-24; Ariel E. Lugo, "Stress and Ecosystems," *Energy and Environmental Stress in Aquatic Systems*, J. H. Thorp and J. W. Gibbons (eds.). DOE Symposium Series (1978), and Kay (1991).

particular view of how to treat a specific place than it is to inform the public of scientific fact.[70]

Ecosystem health. Ecosystem health is similarly an invented attribute of ecosystems and not something included in the concept. It becomes a valid piece of the model only if a management objective is assigned to the manipulation of the ecosystem in question.[71] That is, if one is to talk of ecosystem health, one also must state the objective desired for that ecosystem be it a cornfield, windbreak, pristine mountain lake, or tropical estuary and be it (the objective) food, fiber, or spiritual tranquility. It is possible to design any number of objectives and activities to make use of the attributes of any system. Manipulation of the system to meet an objective, of course, includes the tools required for conservation or preservation as well as for other forms of development. Characteristically, if there is more than one objective for that system at a given time, the result will always be conflict over how "best" to use its structure and function on the one hand and just who is going to make that decision on the other.[72]

Ecosystem Restoration. Likewise, given the dynamics of ecosystem change, *ecosystem restoration* is a difficult term to understand and an even more difficult task to pull off.[73] Succession, a notion put forth by ecologists to describe ecosystem formation, is really a continuum of evolving structure and function altered by forcing functions known and unknown to the restorer. The decision as to where along this continuum one can say an ecosystem is now "restored" is a decision made according to the desires (objectives) of the restorer which may or may not be what the system looked like when it was "in equilibrium" or what it may have been like before humans got there.[74] Indeed, according to the Society for Ecosystem Restoration, ecosystem restoration can be based on conservation goals, economic need, ethical or cultural values, aesthetic principles, or, even, political considerations.[75]

---

70 Try an interesting experiment with reference to science vs. advocacy: Type "ecosystem fragile" and "ecosystem stress" into a search engine and see how many of the references under each can be classified as "advocacy" or "science."

71 Robert T. Lackey, "Values, Policy, and Ecosystem Health." *BioScience* 51.6 (2001) 437-43.

72 OAS, *Minimum Conflict: Guidelines for Planning the Use of American Humid Tropic Environments.* (Washington D. C.: Organization of American States, 1987) 198 pp.

73 James G. Wyant, Richard A. Meganck, and Sam H. Ham, "The need for an environmental restoration decision framework." *Ecological Engineering* 5 (1995) 417-20.

74 Kay (1991).

75 SER, "Global Rationale for Ecological Restoration" IUCN-CEM 2nd Ecosystem Restoration Working Group Meeting, 2-5 March 2003. Taman Negara, Malaysia.

Sustainable development

Because of its acceptance across a broad span of sectors and the numerous problems that the phrase "sustainable development" creates in terms of definition and practical application, its handling within GEG falls into the category of positive feedback for something less than optimal.

The abstract and the practical are two levels of conceptual intensity that fall out of the hundred or so definitions of sustainable development. All of us are captivated by, and intuitively understand, the abstract idea of sustainable development. What we grasp of it at a practical level, is something else altogether. Being understood differently by each, sustainable development is a term that most politicians, entrepreneurs, and environmentalists alike, enthusiastically embrace; development *and* conservation together are a heady mix. Add to that the positive feedback every time anyone uses the term, no matter the context and content, and we have the stuff of conflict and chaos.

This does not mean that we must rid ourselves of the term by any means. The idea is certainly a giant step forward from unsustainable development and every initiative that takes us forward is an improvement on what we have. However, GEG could make more progress if the supporters of sustainable development would accept that the concept has both abstract and practical levels. Then, believing the abstract to be possible, they could dedicate themselves to work through the difficulties and contradictions of development rather than denying that such contradictions exist just because someone adds the word "sustainable" to the project or policy.[76] The alternatives, the ones based on what we would like to know rather than on what we do know; or the ones that respond to pressures – be they from the despoilers, or from the conservationists – lead us only to unproductive chaos.

---

76 All development decision making, "traditional" as well as "sustainable," suffers from a "failure to confront the inherent conflicts between the technical, political and ethical rationalities" of the process where a decision that favors any one of the three very often conflicts with one or both of the other two (See Denis Goulet, "Three Rationalities in Development Decision-making." *World Development*, 14.2 (1986) 301-17. Further, the three principles of sustainable development most often cited (protection of nature for its own sake, efficiency in the use of resources, and intergenerational equity) are mutually exclusive ideas according to David J. Pannell and Steven Schilizzi, "Sustainable Agriculture: A Question of Ecology, Equity, Economic Efficiency or Expedience?" *Sustainability and Economics in Agriculture*. SEA Working Paper 97/1 (1997) GRDC Project, University of Western Australia. Working through these problems is the fundamental challenge for both environmental management and global environmental governance.

C.H.A.O.S.S. on Chaos

Complexity

According to W. Brian Arthur, who helped formalize Chaos Theory:[77]

> [complexity] tends to increase as functions and modifications are added to a system to break through limitations, handle exceptional circumstances or adapt to a world itself more complex... Where forces exist to weed out useless functions, increased complexity delivers a smooth, efficient machine. Where they do not, it merely encumbers.[78]

This statement by Dr. Arthur is one that the practitioners of GEG need to heed as they try to "break through limitations, handle exceptional circumstances and adapt to a more complex world." For example, based on the conflict seen during both the run-up and the WSSD event itself, the process of GEG is neither "smooth" nor "efficient."

Some of the reasons for this lack of smoothness and efficiency, already mentioned, consist of the divergent policies, ideologies, social and economic conditions and cultures of nearly two hundred independent states and their governments as well as the policies, ideologies and cultures of thousands of NGOs. Likewise, the broad and changing makeup of each of the governments and non-governmental organizations tend to promote disappointment rather than firm progress in their search for the means to govern well. Finding a route through and around these permanent fixtures on the GEG landscape requires a concerted effort on the part of scholars and diplomats. One part of that effort, we believe, is to improve the language and the manner in which it is continuously assembled around the process, architecture, and implementation of GEG because it is here that many of the "useless functions" may be most easily weeded out. One of these encumbrances, in our view, is the term "political will."

Political Will

As previously stated, Chaos Theory includes the idea of "attractors." An attractor is the mathematical conceptualization of some stable point in a complex system. These are the "organizers" of chaos, i.e. when a complex

77 See Chapter I of Waldrop 1992.

78 W. Brian Arthur, "Why Do Things Become More Complex?" *Scientific American* (May 1993) 144.

system is on the "edge of chaos" the influence of an attractor may push (or drag) the system to a new order that can include death – the ultimate in stability, or endless repetition – only slightly more desirable than death. Certain systems also exhibit behaviors that settle on "strange," or chaotic, attractors. Hooked to a strange attractor, the behavior of the system varies unpredictably and spends its life in chaos. In GEG, our guess is that the term "political will" – something that GEG has yet to define or even explain – is a strange attractor.

How obsessed is GEG with the idea of political will? An Internet search for a combined "political will" and "environment" yields 125,000 references; a search for "political will" and "sustainable development," yields 59,000 references. That is a lot of cyberspace dedicated to something that few have ever tried to explain. Even more unsettling, however, is that the majority of these references relate to discussions held during meetings dedicated to GEG and pretty much all of them include what can only be described as an excuse that pushes the blame for failure or inaction onto someone else: "The effort we spent so much time planning, failed because of a lack of political will."

However, to say that a project, program, or policy failed because of a "lack of political will," is little more than pretext to escape the guilt for something we did not do: we did not finish the assignment and our homework was handed in half done. Is it not somewhat embarrassing that we would spend millions upon millions of dollars to develop action proposals whose objectives are nothing less than the salvation of the world as we know it and then not do our homework?

To make some sense of the issue, we at least need a definition of what "political will" is. To start, we will make a leap of faith and say that "political will" involves a political decision. The question then becomes, "How are political decisions made?" Since a political decision is probably not much different from any of the other decisions we make – only the names and intensities change, we have decided that we make decisions based on the route of least conflict for ourselves, the decision maker.

"Ha!" some of you will say, "I knew it all the time. Politicians are such cowards!" Take it a step further, though, and ask "conflict with what or with whom" and "how strong" and "how much" and the whole thing gets a good deal more complicated; so complicated in fact that we find ourselves admiring the politicians who have to make political decisions.

Making a decision requires deliberation. Deliberation is required because one has to balance the amount and kinds of conflict the decision will cause with a number of different entities or ideas, pretty much all of them important. Here are a few: constituents, support base, mentors, peers, opposition, leaders, law, constitution, other priorities, scientific information, aspirations and conscience. Multiply all of that by the many individuals who will become involved in a decision on GEG and "political will" becomes a strange attractor. Thus, relying on political will without doing our homework is costly, dangerous and naïve. However, since progress in GEG does indeed depend on the decisions of politicians, we need to know just what the homework is that went missing.

The answer to that homework question is easy; doing it is quite a bit more difficult. That is, our homework is to discover and manage the conflict that is at the center of emerging change; something few of us are very good at doing. We have difficulty in managing the many conflicting demands made by humans on the systems we share because the competing demands are not discovered early enough, they are known but not included in the analysis and treatment, or, more likely, there is no *real* consensus gathered around the process, the product, or both, before a governance proposal is put forward.[79] The reactions of unsatisfied or disenfranchised affected parties (including a significant portion of the political establishment) are major reasons for failed GEG proposals and actions. They fail because of unresolved conflict.

Emergence

There are at least two ways to manage emerging change in complex social systems such as GEG; they include the participation of civil society and the creative management of conflict.[80] These are important enough that there ought to be whole divisions of GEG practitioners schooled in these two methods for minimizing conflict, the reason being that if we can offer a program or policy to a decision maker that meets a desired objective and minimizes the conflict that the decision will bring, we will have done our homework.

---

79 Consensus, of course, is an important goal of negotiations meant to lead to a binding treaty based on an agreement to use the words that are used. Unfortunately, this does not necessarily include agreement on their definitions.

80 There are others, all of which tend to fit into these two responses. Gender issues, for example, are important, as are issues surrounding youth, indigenous peoples, etc. Including these interests in the analysis reduces future conflict – and produces a better product. Likewise, if one asks of any environmental impact, "Who created it?" and "Who felt it?" one immediately realizes that the "impact" is really a conflict between two or more interests that can be treated. That is something far different from identifying an impact of a development activity on something called *the environment* – a problem that cannot be treated.

Participation of Civil Society. Based on Principle 10 of WCED, two regions have established formal mechanisms to involve civil society in government decision-making processes: the Aårhus Convention, in Europe,[81] and the *Inter-American Strategy for the Promotion of Public Participation in Decision-Making for Sustainable Development* (ISP) in the Americas.[82] Both instruments include a set of principles that help guarantee civil society input into decision-making. These mechanisms, combined with the growing activism of transnational elements of civil society, are radically transforming international negotiation, diplomacy, and governance.

Europe's "Aårhus Convention" (The United Nations Economic Commission for Europe Convention on Access to Information, Public Participation in Decision-making and Access to Justice in Environmental Matters) was adopted in June 1998 and entered into force in October of 2002. Not only was it the first formal agreement of its type in the world, it represents a departure from the standard convention in that it links environmental rights and human rights and acknowledges the obligation owed to both present and future generations.[83] It establishes that sustainable development can be achieved only through the involvement of all stakeholders and it links government accountability and environmental management. The Convention focuses on interactions between the public and public authorities in a democratic context and it is forging a new process for public participation in the negotiation and implementation of international agreements.

The subject matter of the Aårhus Convention also goes to the heart of the relationship between people and their governments. It is not just a convention concerned with environmental management; it is also an agreement that demands government accountability, transparency, and responsiveness. The Aårhus Convention grants the public rights and imposes obligations on parties and public authorities regarding access to information, to justice, and to the process itself.

---

81 UNECE, *Introducing the Aårhus Convention*. Accessed 15 August 2001 at http://www.buwal.ch/inter/e/ea_zugan.htm.

82 OAS, *Inter-American Strategy for Public Participation in Decision-Making for Sustainable Development*. (Washington, D. C.: General Secretariat, Organization of American States, 2001).

83 The XXXII General Assembly of the OAS (Barbados, 2002) also followed this path in passing Resolution 1896 which clearly joins the right to environmental goods and services, such as clean air and water, to human rights via the Inter-American Human Rights Commission.

The *Inter-American Strategy for Public Participation* also has its origins in UNCED. UNCED led to the first hemispheric "Specialized Summit of the Americas on Sustainable Development." This summit, convened in Santa Cruz, Bolivia in late 1996, instructed the Organization of American States to formulate an inter-American strategy to promote public participation in decision-making for sustainable development. In a show of consistency, subsequent summits and meetings of ministers (Chile 1996; Quebec City 2001) endorsed the concept of public participation in government decision-making.

Consequently, early in 1997, the OAS General Secretariat began formulation of the ISP as an open, transparent, and participatory initiative itself. A Project Advisory Committee (PAC) that included members from governments, private business, and labor, as well as women and indigenous people, provided strategic guidance and advice. The core of the Strategy is the Policy Framework that contains the following six basic principles:

- A proactive role of governments and civil society to assure opportunities for public participation;
- Inclusion of a diversity of interests and sectors;
- Responsibility to share the commitments and burdens of development;
- Comprehensiveness to ensure participation in all phases of the decision-making process, with sufficient flexibility to make midstream adaptations;
- Access to relevant information, to the political process, and to the justice system; and,
- Transparency of information within, among, and between government and civil society organizations to ensure the efficient use of resources.

These principles, in turn, gave rise to the following recommendations:

- Create or strengthen existing formal and informal communication mechanisms to encourage information sharing, collaboration, and cooperation among civil society groups, within and between levels of government, and between government and civil society.
- Create, expand, and put into practice legal and regulatory frameworks that allow for the participation of civil society in development decisions.
- Support institutional structures, policies, and procedures that actively promote and facilitate government and civil society interaction in development decisions.

- Build and strengthen the capacity of individuals, within government and civil society organizations, to participate in development decision-making with an increased base of knowledge on sustainable development issues and public participation practices.
- Procure and expand financial resources to initiate, fortify, and/or continue participatory practices in development decision-making. And,
- Create, strengthen, and support formal and informal fora in which development activities are discussed and related decisions taken.

All this, of course, may (will) create any number of temporary problems and will often slow progress considerably. However, the advantages do seem to outweigh the disadvantages and the participation of civil society does represent an emerging solution to which GEG must pay close attention.

Conflict Management. In terms of GEG decision-making, we have come a long way in the years since the 1972 Stockholm Conference but there is still a long way to go. Along the route, we will need to work harder at conflict management – something that is a great deal different from the frequently acrimonious negotiations that currently take place in GEG.

Obviously, conflict management has been around forever and it is something we still need.[84] Unfortunately, however, where GEG is concerned, conflict management is too often seen as compromise and a loss or restriction of power when what it should be seen as is the opportunity to create.[85] Does that sound naïve? Of course it does; at least it does to us. But it also sounds necessary. We do live in a world of inequities in information, finances, organization, abilities, and opportunity as well as hidden agendas and unarticulated, though strongly felt, needs. Human values are as diverse as personality, and all of them are important. Memories are long and history carries more weight than it should. Conflict is everywhere.

However, though conflict is often insane, it also tells us that inequities do exist, that noise makes problems easier to identify, and that there is value in debate. Because of all of this, conflict management is not just a means to resolve disputes. It also is an opportunity to use the relationship revealed by a conflict to engage in productive debate, and to look for new and different

84 William I. Zartman, "Lessons for Analysis and Practice" and Jeffery Z. Rubin, "Third Party Roles: Mediation in International Environmental Disputes," both in Sjostedt (1993).

85 Dudley Weeks, *Eight Essential Steps to Conflict Resolution* (New York: G.P. Putnam's Sons (1992) 290 pp.

solutions to the problems uncovered because of that conflict.
According to Dudley Weeks, who teaches conflict resolution at the American University in Washington, D.C., at least four of our perceptions of conflict need to be changed:[86]

- That "conflict is always a disruption of order", a negative experience when in fact it is an outgrowth of diversity that can be used to clarify a relationship and additional ways of thinking;
- That conflict is always a battle between competing and incompatible self-interests while forgetting the presence of needs or goals that the two [or more] parties might actually share;
- That a conflict defines the entire relationship when the reality is that it is only one part of a complex relationship; and
- That the conflict involves two absolutes, between right and wrong or between good and evil when, many "values" are, in fact, subjective preferences.

Accordingly, conflict management may well be the single most important ingredient for successful global environmental governance that we know of – if we view conflict in positive ways rather than as something that requires the choosing of sides and destructive engagement to the bitter end.

Justice and the problem of Environment vs. Environments

Perhaps some of you have noticed that we, the authors, have had a difficult time in keeping our eyes fixed on what we set out to do in writing this essay: to discuss only the language problem in GEG and leave the problems of power imbalance, cultural differences and all those "non-environmentalists" trying to get into GEG to someone else. We had an idea all along, however, that in the end those two problems would turn out to be essentially the same thing. Since it is something that GEG is going to have to face squarely, we should do it also. It has to do with how one understands the concept of "environment."

The many differences in how the term is used do not appear bothersome within defined communities. For example, all architects recognize that a sleeping environment is different from an eating environment and those, from a work environment. Biologists understand that a fluvial environment is different from a terrestrial environment and

86 Weeks (1992).

any sociologist can tell you the differences between the environments of a *barrio bajo* and a *barrio alto* in Latin America or between Upper and Lower Manhattan in New York City. Few can tell you what "the environment" is, however, and therein lies the problem.

A strong disconnect exists between what many people vaguely believe the term to mean and the reality that those who formulated the concept meant to describe: a system outside of another system that influences the interior system's health, welfare, and behavior. Given that concept, our work is much more than the "preservation of nature," our challenges greater than the "sustainable management of natural resources," and the labor of environmental governance is more than biodiversity conservation, and work on the problems of ozone depletion and climate change – though all of these are extremely important and correctly form a part of what GEG is about.

The concept of environment, however, demands that GEG go beyond pollution control and resource preservation. It obligates us to understand and use the term as a plural rather than as a singular. "Whose environment is to be conserved, whose environment is to be developed, and whose environment will be impacted" by either of those actions are questions asked by the developing countries in every environmental meeting since Stockholm. Indeed, Principle 1 of the Stockholm Declaration affirms that "Man has the fundamental right to freedom, equality and adequate conditions of life, in *an* environment of a quality that permits a life of dignity and well-being" (emphasis added).[87]

Environmental quality is the link between environment and poverty, between environment and health, and between environment and development. It has to do with how well a specific environment satisfies the needs of those who are at its center. Each of us, individually or in small groups, lives in different and ever-changing environments. Some of these environments are rich and rewarding. Others are totally lacking in the means required to satisfy human needs. A singular entity labeled "the environment" does not exist, and calls to protect "the environment" as described by the powerful and the favored disenfranchise the powerless and less fortunate.

Regrettably, the question, "Whose environment are we talking about?" has always received the wrong answer. To respond directly or indirectly

---

87 The text of the Stockholm Declaration can be found at http://www.unep.org/document/default.atp?documentID=978&articleID=1501.

that there is only "the environment," that it belongs to everyone, and that it must be protected, can misrepresent development choices, misdirects human energy and other resources, and hides conflict instead of managing it. Thinking of environment as a singular confounds the issues. Thinking of environment as a plural permits isolation and treatment of problems and allows us to understand the sources of conflict. Use of the plural allows a proper response to the complexity of our universe; use of the singular oversimplifies to an extreme. Use of the singular buries conflict until it explodes and becomes unmanageable. Use of the plural allows civilized discussion, trust and a coordinated search for solutions in the forums of global environmental governance, use of the singular is a major reason why GEG finds itself where it is.

What about "the environment" of planet Earth – the "global environment"? Does it exist? Obviously it does. However, that environment is only one of a countless number and has little to do with the way most individuals or groups perceive the problems found in their own private and local environments.[88] What about the ozone hole and climate change? Are they not environmental problems? Certainly they are. However, rightly or wrongly, the citizens of extreme latitudes are much more concerned about ozone depletion than are those individuals living in regions that are nearer the equator – which just happens to coincide with many of the areas of our world that suffer the most from poverty and disease. Obviously also, climate change will effect different parts of the globe in different ways, some of which may even be positive for a portion of the world's population.[88]

Environment and development are not opposites and the problems of environment and development are not a question of having one or the other in varying degrees. A short comparison will explain: For forest dwellers, deforestation may or may not be an environmental problem that forest protection may or may not solve. Poverty and inequity, on the other hand, entangled as they are with hate and corruption, as well as with international trade, national debt, local incompetence, gender issues, power imbalance, and far from adequate education and health care, are

88 Thomas F. Saarinen, "Environmental Perception." In Ian R. Manners and Marvin W. Mikesell (eds.) *Perspectives on Environment* (Washington D.C.: Association of American Geographers, 1974) 252-89.

89 IPCC. "Climate Change 2001: Impacts, Adaptation and Vulnerability." *IPCC Third Assessment Report* (WMO/UNEP, 2001). See especially par. 1.21 of that report: "Climate Change Represents Opportunities and Risks for Human Development." Accessed at http://www.ipcc.ch.

most certainly environmental problems that only development, and good governance, can unravel.[90] That, in part, is now the future challenge to GEG – a quantum leap in complexity, a long way from the still valid concerns of national parks, wilderness and wildlife – but environment, nonetheless.

---

90 Akiko Domoto, "International Environmental Governance: Its Impact on Social and Human Development." (*Inter-linkages*, World Summit for Sustainable Development. United Nations University Centre. 3-4 September 2001).

A Condensed Glossary of Selected Terms, Concepts, Jargon, Acronyms and Abbreviations Used in Global Environmental Governance

Sources used to Develop the Glossary and List of Acronyms

Much of *C.H.A.O.S.S.* is about words and how they are used in global environmental governance; that is why we have included more than 1,500 definitions and an equal number of abbreviations that represent the devices, fabrications, riddles, acronyms, and jargon invented by our colleagues. We are well aware that this does not do justice to the creativity of GEG wordsmiths. However, having now collected a glossary of some magnitude, we are also aware of the amount of time and work that goes into such an exercise. Because of this, we want to acknowledge the work of all the others who have attempted such a thing and to say thank you for knowingly or unknowingly allowing us to use their work. You will notice that many of the glossary entries end with some hieroglyphics of our own – an attempt to identify our sources and the following is a translation of these hieroglyphics.

We are also quite aware that many readers will disagree with some of the definitions we have chosen to include; we disagree with many of them as well. Let the debate continue; clarification is needed and, for our part, welcome.

ACCU	Asia/Pacific Cultural Centre for UNESCO http://www.accu.or.jp/litdbase/glossary/
AD	Deardorff's Glossary of International Economics. http://www.personal.umich.edu/~alandear
AFESD	Arab Fund for Economic and Social Development. http://www.arabfund.org
AS	Abbott and Snidal (2000).
BCHM	Belgium Clearing House Mechanism. http://webbie.kbinirsnb.be/beh_chd/belgium/glossary/glos-_a.htm.
BLD	Beck's Law Dictionary: A Compendium of International Law Terms and Phrases.
BWP	Bretton Woods Project. http://www.brettonwoodsproject.org/about/glossary.html
CAN	Climate Action Network. http://www.climateactionnetwork.org.
CBD	Convention on Biological Diversity, Secretariat. http://www.biodiv.org/secretariat.

CD	Convention to Combat Desertification. http://www.unccd.int
CF	Canter Fitzgerald. http://www.co2e.com/corporateoveriew.asp
CGG	Commission on Global Governance. http://www.cgg.ch
CIPA	*World Directory of Environmental Organizations. 1996*. Trzyna, T, Margold, E. and J.K. Osborn. California Institute of Public Affairs. 263 p. Sacramento, CA. http://www/cipahq/prg/landmarks.htm.
CITES	Convention on International Trade in Endangered Species of Wild Fauna and Flora. http://www.cites.org.
CIVICUS	CIVICUS. http://www.civicus.org.
CMS	Convention on Migratory Species of Wild Animals. http://www.unep-wcmc.org/cms.
Co	http://www.Chaordic.com.
COICA	Coordinating Group of Amazon Basin Indigenous Organizations. http://www.coica.org.
CONGO	Conference of Non-Governmental Organizations in Consultative Relationship with the United Nations. http://www.ngo.org/congo.html.
CSD	Commission on Sustainable Development. http://www.un.org/esa/sustdev
CSG	Complex Systems Glossary. The Complexity and Artificial Life Research Concept for Self-Organizing Systems. http://www.calresco.org.
Dc	http://www.Dictionary.com
DESIP	Demographic, Environmental, and Security Issues Project. http://www.igc.apc.org/desip/toc.html.
DFID	Department for International Development (UK). http://www.dfid.gov.uk.
DC	Dictionary.Com. http://www.dictionary.com.
EC	Earth Council. http://www.earthcouncil.ac.cr.
Eca	Environment Canada.
ECLAC	United Nations Economic Commission for Latin America and the Caribbean. http://www.eclac.org.
ECO	The Ombudsman Centre for Environment and Development. http://www.omced.org.
eD	eDiplomat. http://www.eDiplomat.com.
EM	Evomarkets. http://www.evomarkets.com/ghg_glossary.html.
ENB	Earth Negotiations Bulletin. http://www.iisd.ca/linkages.
FAO	Food and Agriculture Organization of the United Nations. http://www.fao.org.WAICENT/faoinfo/economics/ESN/codex/default.htm.

FOE	Friends of the Earth International. http://www.foei.org.
GBA	*Global Biodiversity Assessment*. UNEP. 1995. http://www.unep_wcmc.org/assessments.
GBS	*Global Biodiversity Strategy* 1992. WRI, IUCN, UNEP. World Resources Institute. Washington, D.C.
GC	Global Connection.
GCC	Global Climate Coalition http://www.globalclimate.org.
GEF	Global Environment Facility. http://www.gefweb.org/gefgloss.doc.
GP	Greenpeace International. http://www.greenpeace.org.
HH	*Wycoff & Shaw's Harper Handbook* 3rd Ed.
IDB	Inter-American Development Bank. http://www.iadb.org.
IEA	International Energy Agency. http://www.iea.org.
IFAD	International Fund for Agricultural Development. http://www.ifad.org.
IISD	International Institute for Sustainable Development. http://www.iisd1.iisd.ca/glossary.asp.
ILO	International Labor Organization http://www.ilo.org
IMF	International Monetary Fund. http://www.imf.org.
IPPC	Intergovernmental Panel on Climate Change. http://www.ippc.ch.
ITTO	International Tropical Timber Organization. http://www.itto.or.jp.
IUCN	World Conservation Union. http://www.iucn.org.
JG	Joyeeta Gupta. *On Behalf of my Delegation*. http://www.unitar.org/cctrain/Survival%20Guide%20Negotiators%20(nA1P)/www/index.html.
L	Ronnie Lipschutz. http://www.people.ucsc.edu/~rlipsch/publist.htm.
LLL	'Lectric Law Library. http://www.lectlaw.com.
MW	Merriam Webster.
NGS	National Greenhouse Strategy of Australia. http://www.ngs.greenhous.gov.au/glossary.
OAS	Organization of American States, Unit of Sustainable Development and Environment. *Minimum Conflict*. http://www.oas.org/usde.
Ox	Oxfam
PAHO	Pan American Health Organization. http://www.paho.org.
PD	Pavard and Dugdale. (2003)
PIC	Rotterdam Convention. http://www.chem.unep.ch/pic.
PL	*Pequeño Larousse*. http://www.larousse.es/larousse/product.asp?sku=1.

R	Real Alternatives Information Network. http://www.web.net/rain/main.htm.
RC	Ramsar Convention. http://www.ramsar.org.
RCA	Resource Center for the Americas.
Rocky Mountain Institute	http://www.rmi.org/
SNW	Sustainable Northwest http://www.sustainablenorthwest.org
SWF	State of the World Forum. http://www.worldforum.org.
TI	Transparency International. http://www.transparency.org/.
TWN	Third World Network. http://www.twnside.org.sg/title/brie6-cn.htm.
UNCED	United Nations Conference on Environment and Development. http://www.unep.org/unep/partners/un/unced/home.htm.
UNCHS	Habitat. http://www.unchs.org.
UNCLOS	United Nations Convention on the Law of the Sea. http://www.un.org/depts/los.
UNDP	United Nations Development Programme. Governance for sustainable Human Development. http://magnet.undp.org/policy/glossary.htm.
UNED	UNED Forum. http://www.earthsummit2002.org/.
UNEP	United Nations Environment Programme. http://www.unep.org.
UNESCO	United Nations Educational, Scientifical and Cultural Organization. http://www.unesco.org
UNESCO-IHE	UNESCO Institute for Water Education http://www.unesco-ihe
UNFAO	United Nations Food and Agriculture Organization. http://www.fao.org
UNFCCC	United Nations Framework Convention on Climate Change Secretariat. http://www.unfccc.de/siteinfo/glossary.html.
UNFPA	United Nations Population Fund. http://www.unfpa.org.
UNMD	United Nations Millennium Declaration. http://www.un.org.millennium.
UNT	United Nations Treaty Collection, Treaty Reference Guide. http://untreaty.un.org/english/guide/asp.
USGS	United States Geological Survey. Interactive2.usgs.gov/glossary/index.asp.
VC	Vienna Convention on the Law of Treaties. http://www.un.org/law/ilc/texts/cvkengl.html.
W	Whittaker (1983). www.pestmanagement.co.uk/library/glosijk2.html
WB	World Bank. http://www.worldbank.org.

WCMC	World Conservation Monitoring Centre. http://www.unep-wcmc.org.
WEF	World Economic Forum. http://www.weforum.org.
WHO	World Health Organization. http://www.who.int.
WTO	World Trade Organization. http://www.wto.org.
WR	World Reference.com. http://www.worldreference.com.
WRI	World Resources Institute. http://www.wri.org.
WSF	World Social Forum. http://www.forumsocialmundial.org.br.
Wuppertal Institute	http://www.wupperinst.org/sites/links.html
WWF	Worldwide Fund for Nature. http://www.wwf.org.

Glossary

A

A posteriori – Relating to or involving inductive reasoning from particular facts or effects to a general principle; derived from or requiring evidence for its validation or support; empirical; open to revision. (WR)

A précis – An abstract, abridgment, condensation, digest, synopsis, or a summary of the essential thought of a longer piece. (HH)

A priori – Relating to or involving deductive reasoning from a general principle to the expected facts or effects; known to be true independently of or in advance of experience of the subject matter; requiring no evidence for its validation or support. (WR)

Aårhus Convention – The Convention on Access to Information, Public Participation in Decision-making and Access to Justice in Environmental Matters of the UNECE entered into force on 30 October 2001. Its purpose is to link environmental rights to human rights through stakeholder involvement and establishes that the achievement of sustainable development is through government accountability linked to environmental protection. (UNECE)

Aårhus Protocol on Persistent Organic Pollutants (POPS) – A protocol to the Convention on Long Range Transboundary Air Pollution designed to reduce air pollution from 16 persistent organic chemicals including DDT, aldrin, dieldrin, PCB, and the industrial by-products, dioxins and furans. Thirty-three governments signed the protocol in 1998 but it has not yet entered into force. (UNECE)

Abatement – The reduction of the degree or intensity of pollutants or emissions. (EM)

Acceptance – A decision having the same legal affect as ratification and consequently expresses the consent of a state to be bound by a treaty once the minimum number of states has signed and ratified. (VC)

Accession (1) – A sample of a crop variety collected at a specific location and time. (GBS)

Accession (2) – The act whereby a state accepts the offer or the opportunity to become a party to a treaty already negotiated and signed by other states. It has the same legal effect as ratification. (VC)

Accession Countries – The countries that become party, through signing and ratification, to an agreement or treaty already in force. (W)

Acclimation – A change that occurs in an organism that allows it to tolerate a new environment. (CBD)

Accord – An international agreement originally meant to describe something less than a convention or treaty but now all three terms are generally considered as synonymous. (UNT)

Accountability – The requirement that officials answer to stakeholders on the disposal of their powers, duties and decisions; act on criticisms or requirements made of them; and accept the responsibility for their failure, incompetence, or deceit. (UNDP)

Accreditation – A formal registration process that enables representatives of an NGO to attend meetings of an intergovernmental organization after providing information regarding its organization, objectives, and work program, and after its acceptance by the international organization.

Activities implemented jointly (AIJ) – Initiatives taken by countries to jointly implement activities to reduce GHG emissions that, unlike joint implementation (JI), cannot be credited against current commitments of Parties to the UNFCCC. (UNFCCC)

***Ad hoc* group** – A group formed for a specific purpose and no other, which is generally disbanded when the task is completed. (PL)

Ad referendum – An agreement reached by negotiators subject to concurrence by their governments.

Adaptation (1) – Special traits that help a plant or an animal survive in its environment. (CSG)

Adaptation (2) – Policies and actions that minimize adverse impacts of climate change. (UNFCCC)

Adaptation Fund – A fund established to support concrete adaptation projects and programs in developing countries that are Parties to the Kyoto Protocol and that is financed through a share of proceeds from project activities of the clean development mechanism and other sources. (UNFCCC)

Adaptation measures – Actions in response to, or in anticipation of, climate change made to reduce or avoid adverse consequences or to take advantage of beneficial changes. (NGS)

Adaptive management – A type of natural resource management where adjustments are made in response to project monitoring, new scientific information, and changing social conditions that may indicate the need to change a course of action.

Additionality – An intervention that adds to, rather than replaces, other financial inputs. (DFID)

Adoption – The formal act that establishes the structure and content of a proposed treaty text that takes place through the expression of the consent of the states participating in the treat-making process. However, adoption does not yet mean that a State consents to be bound by a treaty. (VC)

Advanced informed agreement – The principle that international exchange of transgenic plants and microorganisms that could adversely affect plants should not proceed (a) without the informed agreement of, or (b) contrary to the decision of, the competent authority in the recipient country. (BCHM)

Adverse effects of climate change – Changes in the physical environment or biota resulting from climate change which have significant deleterious effects on the composition, resilience or productivity of natural and managed ecosystems or on the operation of socio-economic systems or on human health and welfare. (UNFCCC)

Aerosols – Non-gaseous microscopic particles and droplets (.01 - .00001 cm), other than water or ice, suspended in the atmosphere – the most

A

abundant being particles of mineral dust, sulphuric acid, ammonium sulphate, pollens, and carbon or soot. (UNFCCC)

Affordable safe minimum standard approach – A management approach in which the fewest activities required to avoid damages due to climate change require a balance between the acceptance of risk and the affordability of abatement costs.

Afforestation – The establishment of forest cover on land not previously forested. (UNFCCC)

African Development Bank (AfDB or ADB) – The regional development bank for Africa established in 1963. AfDB headquarters is located in Abidjan, Côte d'Ivore and has a membership of 51 African and 26 non-African member states. (CIPA)

African Union (AU) – A pan African organization formed in July of 2002 in Durban, South Africa to replace the Organization of African Unity (OAU). The AU has a security council, a legislature and an economic development plan, called the New Partnership for Africa's Development. (NEPAD)

Agenda – A formally adopted program of work.

Agenda 21 – A blueprint of 40 chapters the fulfillment of which would help lead to the sustainability of development. Nearly 180 governments at the Earth Summit in Rio de Janeiro in 1992 agreed to *Agenda 21*. (UNCED)

Agreements – Legal instruments that are generally less formal and more narrowly focused in subject matter than are treaties. (MW)

Aid – The transfer of goods and services between nations, commercial banks, international agencies or nongovernmental agencies generally in the form of grants and loans. Internationally, aid refers to net flows of official development assistance (ODA for countries on Part I of the DAC list) and official assistance (OA for countries on Part II of the DAC list). (DFID)

Aide mémoire – A signed and dated summary of the points made in an official conversation left with the parties to the conversation, either at the time of the conversation or subsequently, as an aid to memory. (eD)

Alien species – A species occurring in an area outside of its historically known range as a result of intentional or accidental dispersal by human activities. (GBS)

Alliance of Small Island States (AOSIS) – The coalition of over 40 Pacific and Caribbean nations that lobby for action by developed nations to reduce greenhouse gas emissions because of the threat of rising sea levels and increased storm activity which may accompany global warming.

Alternative energy – Energy derived from non-fossil fuel sources. (IPPC)

Ambassador – A diplomatic envoy of the highest rank accredited to a government, generally for a specific period or assignment. (MW)

Ambassador-at large – A Minister of the highest rank not accredited to any particular government, but rather to an organization for a specific task or theme. (MW)

Ambassadress – A female ambassador. (MW)

Amendment – A proposed or effected alteration to a document made by parliamentary procedure.

Annex B Countries – Developed country Parties to the Kyoto Protocol that have agreed to a target for their GHG emissions (generally includes the OECD member states, Central and Eastern Europe, and the Russian Federation. (UNFCCC)

Annex I Parties – Parties to the United Nations Framework Convention on Climate Change who pledged to reduce their greenhouse gas emissions by the year 2000 to 1990 levels. Annex I Parties consist of the Organization for Economic Cooperation and Development (OECD) member states and industrialized countries with economies-in-transition (Central, and Eastern Europe excluding the former Yugoslavia and Albania). (UNFCCC)

Annex II Parties – The developed country Parties and other Parties included in Annex I of the UNFCCC that are also Parties to the Kyoto Protocol.

Annex III Parties – Developing country Parties with economies that are highly dependent on the exploitation, production, processing, and exportation of fossil fuels. (UNFCCC)

Annotated agenda – A draft agenda that includes explanatory text.

Anthropogenic – A thing or phenomenon made by people or that result from human activities.

AOSIS – See the **Alliance of Small Island States**.

Approach – A way of thinking or of doing things based on values used to attack a problem.

Approval – See Acceptance.

Arab Fund for Economic and Social Development – A financially autonomous Pan-Arab regional organization comprising all Arab countries that are members of the League of Arab States. Its functions are to assist the economical and social development of Arab countries through (1) financing development projects, with preference given to overall Arab development and to joint Arab projects; (2) encouraging the investment of private and public funds in Arab projects; and, (3) providing technical assistance services for Arab economic and social development. (AFESD)

Artesian fisheries – Small-scale fisheries carried out by people who rely on fishing to support their families and other local people. (AMNH)

Article 8(j), Working Group on – A working group mandated by COP/CBD to provide advice on the application and development of legal and other appropriate forms of protection for the knowledge of indigenous and local communities embodying traditional lifestyles relevant to the conservation and sustainable use of biodiversity. (ENB)

Aruba Protocol – The protocol concerning Pollution from Land-based Sources and Activities in the Wider Caribbean Region (Oranjestad, 1999) is one of three protocols to the 1983 Cartagena Convention for the Protection and Development of the Marine Environment of the Wider Caribbean Region. (CBD)

Asian Brown Cloud – The haze of pollutants trapped by a temperature inversion over the northern Indian Ocean, much of South Asia, India, Pakistan, Southeast Asia and China that occurs for three to four months each winter and which could have serious effects on the human health and climate of the region.

Asian Development Bank (ADB) – An international development finance institution established in 1966 for the Asia-Pacific region. It has an Office of the Environment established in 1987 to facilitate the integration of environmental planning at all levels of Bank operations. ADB headquarters are located in Manila, Philippines. (CIPA)

Assembly – A gathering of individuals or delegations for the purpose of deliberating and approving legislation, declarations, and policy. (MW)

Assessed contribution – Funds which a member state is required to contribute based on a formula applied to all states which are normally dedicated to support the core functions of the organization. (UN)

Assistance – Support for projects that can include technical advice and assistance as well as funds. (DFID)

Associated project – A GEF project that is either physically dependent on another agency's project or whose success could depend upon the implementation of another project. Opposite of a freestanding project. (GEF)

Atmosphere – The envelope of air surrounding the Earth and bound to it by the Earth's gravitational attraction.

Attractor – A point to which a system tends to move, a goal, either deliberate or constrained by system parameters (laws). The three permanent attractor types are fixed point, cyclic and strange (or chaotic). (CSG)

Audit – A systematic retrospective examination of the whole, or part, of a project to measure conformance with the plan. (MW)

Authentic text – The text of a treaty that is established as definitive and where states cannot unilaterally change its provisions. Usual forms of authentications are by signature, signature *ad referendum* or the initialing by the representatives of the states that negotiated the treaty. (VCLT)

Authentication – The procedure whereby the text of a treaty is established as authentic and definitive usually by signature or initialing by representatives of the negotiating states. (UNT)

B

Balance of payments – A set of accounts for a given period, usually a year, which summarizes the financial transactions of the institutions and resident of one country with the institutions and residents of the rest of the world. The set of accounts consists of a current account, which shows expenditures the country made during the period on the purchase of goods and services from abroad, and the revenue derived from the sale of goods and services to the rest of the world; and a capital account, which shows the flows of private and public investment and of other transfers. (DFID)

Bamako Convention on the Ban of the Imports into Africa and the Control of Transboundary Movement and Management of Hazardous Wastes within Africa (1991) – This agreement provides guidelines for the management of hazardous wastes by members states of the Organization of African Unity (now African Union). It was adopted in Bamako, Mali in January 1991 as an alternative to the Basel Convention as OAU member states felt that the ban on movement of hazardous wastes included in the Basel Convention were not strict enough. To date 22 countries have signed and 18 Countries have ratified the Convention. (UNT)

Bankable – A term used primarily by the international technical assistance agencies and development banks to describe a cost-effective project proposal.

Barbados Declaration and Plan of Action – The formal statement produced by the Global Conference on the Sustainable Development of Small Island Developing States (SIDS) held in Bridgetown, Barbados, 25 April-6 May 1994. The Barbados Declaration made a commitment to specific policies, actions, and measures to be taken at the national, regional and international level to enable SIDS to achieve sustainable development. It recognized the special constraints to sustainable development that SIDS face by virtue of their small size and called for the assistance of regional and international entities, both governmental and non-governmental, to engage in a program of action aimed at the sustainable development of the SIDS. (UNDP)

Basal Convention – The Basel Convention on the Control of Transboundary Movements of Hazardous Wastes and Their Disposal entered into force in 1982 with the goal to protect human and environmental health by minimizing hazardous waste production. The convention requires that the production of hazardous wastes is managed using an "integrated life-cycle approach," which involves strict controls from generation to storage, transport, treatment, reuse, recycling, recovery and final disposal. (UNT)

Baseline (1) – Information collected before or at the start of a project to provide a basis for planning and/or monitoring subsequent progress and impact. (WB)

Baseline (2) – The reference point for calculating incremental costs. The GEF funds the difference between the cost of a project undertaken with global environmental objectives in mind and the costs of the same project without global environmental concerns. The baseline is the latter project that yields only national benefits. (GEF)

Basic human rights – The rights outlined in the "United Nations Universal Declaration of Human Rights" (See Appendix 1).

Beijing Declaration – The Beijing Declaration and its Platform of Action are the products of the Fourth World Conference on Women, which was held in Beijing, China from 4 to 15 September 1995. The Platform for Action is an agenda for women's empowerment. It aims to remove all the obstacles to women's active participation in all spheres of public and private life through a full and equal share in economic, social, cultural, and political decision-making. (UNT)

Beneficiaries – The men and women, communities, or organizations expected to benefit from the project or program.(WB)

Benefit cost ratio – A decision criterion that uses the net present value of the benefit related to the net present value of the costs. (WB)

Benefit sharing – One of the three objectives of the Convention on Biological Diversity it is the "fair and equitable sharing of the benefits arising out of the utilization of genetic resources, including by appropriate access to genetic resources and by appropriate transfer of relevant technologies, taking into account all rights over those resources and to technologies, and by appropriate funding." The Convention also encourages "equitable sharing of the benefits arising from the utilization of knowledge, innovations and practices of indigenous and local communities embodying traditional lifestyles relevant for conservation and sustainable use of biological diversity." (CBD)

Berlin Declaration – A global initiative for the promotion of sustainable tourism, signed in Berlin, March 8, 1997.

Berlin Mandate – A negotiated process from COP 1 (1995) of the UNFCCC that dealt with the inadequacy of the measures committed to under the UNFCCC where developed countries were to return their greenhouse gas emissions to 1990 levels by the year 2000. The Berlin Mandate enables Parties to take action for the period beyond 2000, including a strengthening of developed country commitments, through the adoption of a protocol or other legal instruments.

Best Alternative to a Negotiated Settlement (BATNA) – A resistance level in negotiations corresponding to the best option outside the current relationship.

Bifurcation – A point at which a system splits into two alternative behaviors, either being possible, the one actually followed often being indeterminate or unpredictable. (CSG)

Big table – An informal consultation between Finance ministers from eleven African countries and their counterparts from the development cooperation ministries of the Organization for Economic Co-operation and Development (OECD).

Bilateral – Describes instruments or processes that reciprocally affect two nations or parties.

Bilateral treaty – An agreement between two states.

Biodiversity – The totality of genes, species, and ecosystems in a region or the world (GBS).

Biogeochemical cycle – The chemical interactions among the atmosphere, biosphere, hydrosphere, and the lithosphere.

B

Biological corridor – A management area in which methods are used to maintain the migrations and movement of species between areas of natural distribution.

Biological diversity – (Biodiversity) The variability among living organisms from all sources including, *inter alia*, terrestrial, marine and other aquatic ecosystems and the ecological complexes of which they are part: this includes diversity within species, between species and of ecosystems. (CBD)

Biological resources – (Biotic resources) includes genetic resources, organisms or parts thereof, populations, or any other biotic component of ecosystems with actual or potential use or value for humanity. (CBD)

Biomass – All the living matter in a given area. Also the amount of plant and crop material that could be produced in an ecosystem for making biofuels and other raw materials used in industry, for example.

Biome – The complex of living communities maintained by the climate of a region and characterized by a distinctive type of vegetation.

Bioprospecting – Activities that entail the search for economically valuable genetic and biochemical resources from nature. (BCHM)

Bioregions – A territory defined at a given scale by a distinct combination of biological, social, and geographic criteria, rather than by geopolitical considerations.

Biosecurity – Policies and programs designed to prevent the harmful introduction of plant and animal pathogens into agriculture and food production systems.

Biosphere – The terrestrial and atmospheric area occupied by life. (MW)

Biosphere reserve – A classification of land-use established under UNESCO's Man and Biosphere Programme that includes allied areas of increasing protection intended to demonstrate the relationship between conservation and development.

Biota – A term for all of the organisms, including animals, plants, fungi, and microorganisms, found in a given area. (GBS)

Biotechnology – Any technological application that uses biological systems, living organisms, or derivatives thereof, to make or modify products or processes for specific use. (CBD)

Biotic resources – Those components of biodiversity of direct, indirect, or potential use to humanity. (GBS)

Blast fishing – A method of fishing (usually on reefs) that involves the use of dynamite or homemade explosives to harvest large numbers of fish.

Blocking coalition – A group of opposing interests that could prevent a winning coalition from coming into existence or being sustained.

Blue sector – A catchall label for "environmental" projects that cover issues related to oceans.

Black water/issues – See **Water**.

Blue water/issues – See **Water**.

Brown water/issues – See **Water**.

Body (1) – The main part of a text or of an agreement.

Body (2) – The attendees at a conference or meeting.

Bolivia Summit – The 1996 Specialized Summit of the Americas on Sustainable Development held in Santa Cruz, Bolivia in an attempt to prioritize actions called for in *Agenda 21*.

Bonn Convention – Convention on the Conservation of Migratory Species of Wild Animals (1979) which prohibits the commercial capture of some 50 species of migratory animals and which urges Parties to the Convention to conserve and restore the habitats of these species. (UNT)

Bonn Guidelines – Voluntary guidelines that are the result of Decision VI/24 of COP-VI of the Convention on Biological Diversity that deal with access to genetic resources and the fair and equitable sharing of the benefits arising from their utilization. They are meant to assist Parties, Governments, and other stakeholders when establishing legislative, administrative or policy measures on access and benefit-sharing and/or when negotiating contractual arrangements for access and benefit-sharing. (CBD)

Brackets – Square brackets [ ] are used to surround words or phrases that are in dispute or that are alternative wording proposed by a delegation when drafting legal instruments and indicate that no agreement has been reached.

Bretton Woods Conference – An international conference held July 1-22, 1944 at Bretton Woods, New Hampshire. Attended by representatives of 44 countries, the conference established the International Monetary Fund and the International Bank for Reconstruction and Development (World Bank). (BWP)

Brown cloud – See **Asian Brown Cloud**.

Brown sector – A catchall label for "environmental" projects that cover issues related to urban and industrial areas.

Brundtland Report – The report of the World Commission on Environment and Development (WCED): *Our Common Future* (1987). Gro Harlem Brundtland, then Norwegian Prime Minister, chaired the WCED. (WCED)

Bubble – The idea that emissions reductions anywhere within a specific area count towards a common reduction goal – as if a giant bubble were placed over the various sources to contain them in a common area. (EM)

Bubbles – A term used in the negotiation of multilateral instruments to indicate informal consultations on specific topics that take place formal negotiations.

Bucharest Population Conference – The first World Conference on Population held in Bucharest, Romania in 1974, attended by more than 1,200 representatives of 135 nations.

Buenos Aires Plan of Action (BAPA) – A schedule adopted by COP-4 (1998) of the UNFCCC for reaching agreement on the details of how to implement the Kyoto Protocol.

Buffer zone – The region near the border of a protected area designated as a transition zone between the protected area and other areas managed for different objectives.

Bureau – Group of national representatives that acts between sessions according to the mandates given it by the Parties to a convention or the governing council of an institution. The Bureau presides over the work of a general assembly and facilitates agreement among the Parties or members.

Bureau of Oceans and International Environmental and Scientific Affairs (OES) – The US Department of State's focal point for foreign policy formulation and implementation in global environment, science, and technology issues. (CIPA)

Bushmeat – Edible meat obtained from wild animals. (CBD)

Business Councils for Sustainable Development – (See **World Business Council for Sustainable Development**).

By-catch – The portion of a fishing catch that is discarded as unwanted or commercially unusable.

C

Cairo Plan – Recommendations for stabilizing world population agreed upon at the U.N. International Conference on Population and Development, held in Cairo in September 1994. The plan calls for improved health care and family planning services for women, children and families throughout the world, and also emphasizes the importance of education for girls as a factor in the shift to smaller families. (UNT)

Canadian International Development Agency (CIDA) – The organization through which the federal government of Canada distributes foreign aid.

Canalization – The restriction of possible state space exploration by constraints imposed upon the system either from outside or self-generated, i.e. unavailable possibilities. This helps to preserve stability or the "status-quo" but may also prevent better optima from being reached. (CSG)

Capacity building – A collection of activities including human resources development through education and training, institutional and infrastructure development, and formulation of enabling policies (WCED). A coordinated process of deliberate interventions by insiders and/or outsiders of a given society leading to (i) skill upgrading, both general and specific, (ii) procedural improvements, and (iii) organizational strengthening. (WB)

Capacity development – The process by which individuals, organizations, institutions and societies develop their abilities individually and collectively to perform functions, solve problems and set and achieve objectives. (UNDP)

Captive breeding programs – The founding and management of captive populations of species at risk based upon sound scientific principles for the purpose of securing their survival through stable, self-sustaining populations. (IUCN)

Carbon cycle – The reservoirs of carbon (atmosphere, terrestrial and marine biomass, soil organic matter, fossil fuels and other sediments, the ocean, atmosphere etc.) and the chemical processes governing the exchanges of carbon between them.

Carbon dioxide equivalents – A metric measure used to compare the emissions from various greenhouse gases based upon their global warming potential (GWP) commonly expressed as "million metric tons of carbon dioxide equivalents (MMTCDE)" or "million short tons of carbon dioxide equivalents (MSTCDE)". Multiplying the tons of the gas by the associated GWP derives the carbon dioxide equivalent for a gas. (EEA)

Carbon dioxide fertilization – A process leading to the enrichment of the atmosphere with CO_2.

Carbon sequestration – The uptake and storage of carbon, including anthropogenic carbon dioxide that would otherwise affect global climate change, in large-scale carbon sinks or reservoirs that absorb more carbon than they release (i.e. oceans and immature forests).

Carbon sink – Forests and other ecosystems that absorb carbon, thereby removing it from the atmosphere and offsetting CO_2 emissions. The Kyoto Protocol allows certain human-induced sinks activities undertaken since 1990 to be counted towards Annex I Parties' emission targets. (EEA)

Caring for the Earth – The 1991 sequel to the *World Conservation Strategy* (published in 1980). Like the earlier WCS, *Caring for the Earth* was sponsored by the IUCN, UNEP and WWF. It sets forth 132 actions required to "increase human well-being and halt the destruction of the Earth's capacity to support life." (CIPA)

Carrying capacity – A term used to describe the maximum number of individuals of a particular species that a given environment can support over time and within set management objectives and methods.

Cartagena Convention – The Convention for the Protection and Development of the Marine Environment of the Wider Caribbean Region, adopted in Cartagena, Colombia on 24 March 1983 and entered into force on 11 October 1986 as the legal instrument for implementation of the Caribbean Action Plan. (UNEP-CAR-RCU)

Cartagena Protocol – A protocol of the CBD on biosafety developed at an

C

extraordinary Conference of the Parties in Cartagena, Colombia in 1999 and adopted by more than 130 countries on January 29, 2000, in Montreal, Canada. It provides a framework for addressing environmental impacts of bio-engineered products (referred to as living modified organisms or "LMOs") that cross international borders. (CBD)

Catchment area – A river or lake drainage basin. (UNDP)

Central Group-11 (CG-11) – A coalition of governments that brings together the countries with economies in transition listed in Annex I of the UNFCCC minus the Russian Federation and the Ukraine (UNFCCC)

Centre of crop diversity – Geographic area containing a high level of genetic diversity for crop species in *in-situ* conditions. (CBD)

Centre of origin – Geographical area where a plant species, either domesticated or wild, first developed its distinctive properties. (CBD)

Certification – A written testimony to certain facts used to satisfy regulatory bodies or customers that a person's or thing's capability, qualities, performance, etc. meet specified standards.

Certified emission reductions (CERs) – Verified and authenticated reductions of greenhouse gas from the abatement or sequestration projects that are certified by the Clean Development Mechanism. (UNFCCC)

Chain of custody – A term used in certification that tracks all stages of the production, handling, distribution and sale of the product to be certified to insure that the product meets the standards set by the certifying institution.

Chair – The title of the person charged with facilitating work of a meeting, and for ensuring that the rules of procedure for that meeting are upheld.

Chaord – Any auto catalytic, self-regulating, adaptive, nonlinear, complex organism, organization, or system, whether physical, biological or social, the behavior of which harmoniously exhibits characteristics of both order and chaos. (Co)

Chaos – A system whose long-term behavior is unpredictable, tiny changes in the accuracy of the starting value rapidly diverge to anywhere in its possible state space. There can, however, be a finite number of available states, so statistical prediction can still be useful. (CSG)

Chaos Theory – An attempt to uncover the statistical regularity hidden in processes that otherwise appear random. Systems described as "chaotic" are extremely susceptible to changes in initial conditions. As a result, small uncertainties are magnified over time, making chaotic systems predictable in principle but unpredictable in practice.

Chapeau – An opening paragraph to a statement or declaration used to outline or introduce the topic at hand rather than to directly enter into the discussion.

Charrette – A short period of intense, cross-disciplinary, collaborative planning that empowers community participants and thereby encourages their support for, and ownership of, the final product.

Charter – A term used for particularly formal and solemn instruments, such as the constituent treaty of an international organization. (UNT)

Chilling effect – A situation where compliance with trade rules proves to be a barrier to the use of trade provisions to stop environmentally damaging activities (i.e. POPs: Biosafety Protocol intellectual property debate).

Chlorofluorocarbons (CFCs) – Chemicals used for refrigeration, air conditioning, packaging, insulation, solvents, or aerosol propellants that are GHGs and that, under certain conditions break down ozone.

CIVICUS – An international alliance dedicated to strengthening citizen action and civil society throughout the world. Its members include nearly 500 nongovernmental organizations, associations, private foundations, and individuals from 91 countries. It works to promote an enabling architecture for civil society; promote citizen participation in civil society; and build a global civil-society movement. (CIVICUS)

Civil society – Extra-governmental individuals and groups, organized or unorganized, who interact in the social, political and economic domains and who are regulated by formal and informal rules and laws. Civil society

C

offers a dynamic, multi-layered wealth of perspectives and values, seeking expression in the public sphere (UNDP). A classification that includes those political, cultural, and social organizations of modern societies that are autonomous of the state but part of the mutually constitutive relationship between state and society (L). The web of associations, social norms and practices that comprise activities of a society as separate from its state and market institutions. Civil society includes religious organizations, foundations, guilds, professional associations, labor unions, academic institutions, media, pressure groups and political parties. (WB)

Classical statist approach to international law – A view of international law wherein states are the only recognized and legitimate "subjects," everything else is an "object," and only states can create and employ international law. It is premised on ideas regarding the "sovereign equality of states," a duty of non-intervention on the part of states in the internal affairs of other states, and state consent to international obligations. (ECO)

Clean Development Mechanism (CDM) – CDM is a means by which non Annex I Parties to the Kyoto Protocol (developing countries) can be assisted in development efforts and contribute to the stabilization of GHG concentration in the atmosphere at a level that would prevent dangerous anthropogenic interference with the climate system and to help developed countries and countries with economies in transition to meet their quantified emission limitation commitment. (UNFCCC)

Cleaner production – The continuous application of an integrated preventive environmental strategy to processes, products, and services to increase overall efficiency, and to reduce risks to humans and the environment. Cleaner production can be applied to the processes used in any industry, to products themselves, and to various services provided in society. (UNEP)

Clearing House Mechanism (CHM) – A mechanism generally operated by a convention's secretariat to promote and facilitate technical and scientific cooperation or to facilitate the exchange of scientific, technical, and legal information and assist developing country Parties in the implementation of the MEA concerned. (UNFCC)

Climate Action Network (CAN) – Formed in 1989, CAN is a global network of over 287 Non- Governmental Organizations (NGOs) working to promote government and individual action to limit human-induced climate change to ecologically sustainable levels. CAN members work to achieve this goal through the coordination of information exchange and NGO strategies on international, regional and national climate issues. (CAN)

Climate change – A change of climate that is attributed directly or indirectly to human activity that alters the composition of the global atmosphere and which is in addition to natural climate variability observed over comparable time periods. (UNFCCC)

Climate models – Large and complex computer programs used to simulate global climate based on mathematical equations derived from the physics that govern the earth atmosphere system. (IPPC)

Climate system – The totality of the atmosphere, hydrosphere, biosphere and geosphere and their interactions. (UNFCCC)

Clustering – A suggested method to reduce conflict and overlap between international legal instruments which includes placing treaties or conventions having similar subject matter or which cover the same geographic area in the same administrative unit.

Coalition – A temporary alliance of distinct parties, persons, or states that is generally established to further joint action on a specific matter.

Cocoyoc Declaration – The published report of the 1974 UNEP/UNCTAD symposium held in Cocoyoc Mexico that identified the maldistribution of resources as a factor in environmental degradation. The symposium, chaired by Barbara Ward, called for development action focused on filling basic human needs. (UNT)

Codex Alimentarius – The international "food code" or set of standards implemented through a commission responsible to the Directors-General of the Food and Agriculture Organization (FAO) and the World Health Organization (WHO). The standards are established to protect the health of consumers and ensure fair practices in the food trade; promote coordination of all work on food standards undertaken by international, governmental and non-governmental organizations; and to determine priorities and guide the preparation of draft standards. (FAO)

C

Co-financing – An arrangement whereby an international bank attracts a second lender to finance part of a loan. Such an arrangement spreads the risks to the banks and obtains a second opinion as to whether the project is bankable and interesting. (DFID)

Co-generation – The use of waste heat from electric generation, such as exhaust from gas turbines, for either industrial purposes of district heating. (IPCC) Also, the use of a public power grid by private energy producers.
Collaboration – Cooperation with an agency or instrumentality with which one is not immediately connected. (MW)

Collective decision-making process – The negotiation of international agreements where the established procedures include decision criteria of fair representation and treatment of all parties, transparency, equity and efficiency.

Comity – The agreement between countries to recognize and respect the laws and institutions of each. Somewhat similar to international law except that international law is binding while comity is not. (MW)

Command and control – In relation to policy and management, command-and-control instruments (e.g. mechanisms, laws, measures) rely on prescribing rules and standards and using sanctions to enforce compliance with them. (EEA)

Commission on Global Governance (CGG) – An independent group of 28 public figures who, based on the belief that the end of the Cold War offered opportunities to build a more cooperative, safer and fairer world, published *Our Global Neighborhood* (1995) and *Millennium Year and the Reform Process* (2000).

Commission on Sustainable Development (CSD) – A commission of the UN Economic and Social Council created in December 1992 to ensure effective follow-up of UNCED; and to monitor and report on implementation of the Earth Summit and the WSSD agreements at the local, national, regional and international levels. (CSD)

Commitment authority – The power that GEF has to make commitments based on the amount of resources it has received or which donors have committed. The authority covers contributions received in the form of notes and cash deposits, and investment income generated on cash deposits, less actual project disbursements and anticipated administrative costs of the GEF. (GEF)

Committee – A group of people selected from a larger group to discuss, investigate or report on a particular subject.

Committee of the whole (COW) – A committee created by a COP to facilitate the process of text negotiation. The COW the turns over the draft text to the COP for further work and formal adoption during a plenary session. (BCHM)

Common but differentiated responsibilities – Principle 7 of the Rio Declaration which reads, "[S]tates shall cooperate in a spirit of global partnership to conserve, protect and restore the health and integrity of the Earth's ecosystem. In view of the different contributions to global environmental degradation, States have common but differentiated responsibilities."

Common concern of humankind – A principle supported by numerous international meetings that represents an effort to provide a basis for international action to protect the global climate. However, since climate change is not imposed by one state upon another state, the traditional legal principles governing transboundary pollution do not apply. But if the atmosphere is a "common concern of humankind," all states have an interest and duty to protect it from serious harm. A state on one side of the globe is thus "affected" by a state on the other side of globe that is emitting greenhouse gases into the atmosphere. (UNFCCC)

Common good – Freely accessible good for which there is strong competition between users. This is the case with fishery resources, certain forest areas and certain grazing rights without an agreement, unrestricted exploitation of the resource may lead to it becoming exhausted.

Common heritage of mankind – A concept initially used by Arvid Prado in 1967 which stated that the seabed and the ocean floor (and presumably other of the global commons) are beyond national jurisdiction are the common heritage of mankind, not subject to national appropriation in any manner whatsoever and should be used and exploited for peaceful purposes and for the exclusive benefit of mankind as a whole. (UNCLOS)

C

Common property resource – A resource where the rights of use are communally shared and owned and where formal or informal use regulations are employed.

Community – An integrated group of species inhabiting a given area. (GBS)

Community-based organizations (CBO) – Organizations that are voluntary, not-for-profit, non-governmental and run by citizens with aims and activities limited to a locality that has a unique identity, as a city or neighborhood.

Community of practice (CoP) – A group of people informally bound together by shared expertise and passion for a joint enterprise in a common context. (Wenger)

Competition – The idea that to survive, agents must fight each other and that only one of them can be successful. This assumes that resources are limited (insufficient for both) and is often a negative-sum strategy, i.e. "win-lose" or "lose-lose". (CSG)

Complex – Intricate association of individual parts forming a whole.

Complex adaptive system – A form of system containing many autonomous agents who self-organize in a coevolutionary way to optimize their separate values. (CSG)

Complex systems – Systems that interact non-linearly with their environment and their components have properties of self-organization that make them non-predictable beyond a certain temporal window. (PD)

Complexity theory – A field of research that explores systems in which a great many independent agents interact with each other in a great many ways. (W)

Concessional financing – Usually, money loaned by development banks at below market interest rates and often below the cost of borrowing for the lending agency to promote a specific goal.

C

Concessional loan – Loan provided to poorest countries with lower interest rates and longer repayment periods than typical of standard market or multilateral loans, i.e. less than market interest rates and extended grace period.

Conciliation – A term meaning to unite or to make compatible different viewpoints in a document.

Conditionality (1) – The practice of attaching conditions to international economic concessions or development assistance funds, or to the opening of international markets. The obligations accepted by a developing country in relation to its policies or institutions as part of an aid transaction. (DFID)

Conditionality (2) – A rule of the International Monetary Fund that a member's right to the use of credit tranches and credit facilities will be conditioned on its progress in regularizing its balance-of-payments obligations and in developing sustained economic growth.

Conditions – The existence of, or setting of, certain requirements before agreement can be reached or work can begin.

Conference (or Congress) – International meeting. In the diplomatic sense, a congress has the same meaning as a conference. (eD)

Conference of Parties (COP) – The supreme body of any multilateral environmental agreement (MEA). The role of the COP is to promote the MEA and its objectives and to review its commitments, effectiveness and implementation. COPs generally meet once a year to review implementation and take decisions on how to improve the implementation process.

Conference-room papers (CRP) – A category of in-session documents containing new proposals or outcomes of in-session work for use only during the session. (BCHM)

Conflict – A condition where two or more parties having differing interests or perspectives that require resolution to achieve an end.

Consensus – A decision of an assembly that has agreement from all voting parties or members. An agreement adopted without a vote when there are no stated objections by any of the delegations. Abstentions may be voiced and noted in the conference record and not affect a consensus decision.

C

Consequential equity – A principle of equity that refers to the outcomes of a decision in terms of burden sharing among and within countries as well as between present and future generations.

Conservation – The management of human use of the biosphere so that it may yield the greatest sustainable benefit to current generations while maintaining its potential to meet the needs and aspirations of future generations. (GBS)

Conservation categories of IUCN – Species conservation groupings that indicate the degree of threat to survival for taxa. The six IUCN categories are given a capital letter to distinguish them from general usage of the terms as follows:

- Threatened Categories
  - *Extinct (Ex)* Taxa which are no longer known to exist in the wild after repeated searches of their type localities and other known or likely places.
  - *Endangered (E)* Taxa in danger of extinction and whose survival is unlikely if the causal factors continue operating. Included are taxa whose numbers have been reduced to a critical level or whose habitats have been so drastically reduced that they are deemed to be in immediate danger of extinction.
  - *Vulnerable (V)* Taxa believed likely to move into the Endangered category in the near future if the causal factors continue operating. Included are taxa of which most or all the populations are decreasing because of over-exploitation, extensive habitat or other environmental disturbance; taxa with populations that have been seriously depleted and whose ultimate security is not yet assured; and taxa with populations that are still abundant but are under threat from serious adverse factors throughout their range.
  - *Rare (R)* Taxa with small world populations that are not at present Endangered or Vulnerable, but are at risk. These taxa are usually localized within restricted geographical areas or habitats or are thinly scattered over a more extensive range.
  - *Indeterminate (I)* Taxa known to be Extinct, Endangered, Vulnerable, or Rare but there is not enough information to say which of the four categories is appropriate.

- Unknown Categories
  - *Status Unknown (?)* No information is available with which to assign a conservation category.
  - *Insufficiently Known (K)* Taxa that are suspected but not definitely known to belong to any of the above categories, following assessment, because of lack of information.
- Not Threatened Category
  - *Safe (nt)* Neither rare nor threatened.

Constructive ambiguity – A method used in diplomacy that inserts temporary language into a document that is the lowest common denominator acceptable by the parties in disagreement, but which allows negotiation to proceed – hopefully to full agreement on other more important issues. The wording must be revisited or implementation of the agreement will be made more difficult later.

Consultative group – Groups established to treat specific themes or geographic regions where the members are usually both public and private and where their support may or may not include financing.

Consultative Group on International Agricultural Research (CGIAR) – CGIAR was created in 1971 as an association of public and private members to support a system of 16 Future Harvest Centers that work in more than 100 countries, with a mandate to mobilize cutting-edge science that will reduce hunger and poverty, and improve human nutrition and health, while emphasizing environmental protection. (FAO)

Contact group – A group called into being to resolve a specific issue on which there is disagreement. Membership is ostensibly open to all Parties, but is usually limited to those Parties individually invited by the Chair to participate because of their different viewpoints. (ENB)

Continental shelf – The seabed and subsoil of the submarine areas that extend beyond a coastal state's territorial sea throughout the natural prolongation of its land territory to the outer edge of the continental margin. A coastal state may claim a continental shelf of up to 200 miles from the baselines from which the territorial sea is measured even if the continental margin is not that far seaward; but its maximum claim can be no more than 350 miles.

Contra legem – "Against the law" (term used to describe an equitable decision of a court or tribunal that is contrary to the law governing the controversy. (BLD)

C

Contracting state – A State that has consented to be bound by the treaty, whether or not the treaty has entered into force. (VC)

Convention – When signed and ratified, a binding agreement between two or more states that deals with specific matters of mutual interest and which defines the duties of the Parties to the Convention that sets out the political and legal framework for actions that may be listed or that may need to be developed. (VC)

Convention Concerning the Protection of the World Cultural and Natural Heritage – A convention adopted in November 1972 at the 17th General Conference of UNESCO. The Convention is responsible for the establishment and recording of the World Heritage List of cultural and natural properties, submitted by the States Parties. These sites are considered to be of outstanding universal value. The Convention provides technical co-operation under the World Heritage Fund to States Parties whose resources are insufficient to safeguard World Heritage properties.

Convention for the Protection of the Ozone Layer – A multilateral treaty adopted at Vienna on March 22, 1985, and in force from September 22, 1988. It requires States Parties to cooperate in undertaking scientific research to protect the earth's ozone layer.

Convention on Biological Diversity (CBD) – The Convention on Biological Diversity was negotiated under the auspices of UNEP and opened for signature at UNCED in June 1992. It entered into force on 29 December 1993 and has three goals: to promote the conservation of biodiversity, the sustainable use of its components, and the fair and equitable sharing of benefits arising out of the utilization of genetic resources. (ENB)

Convention on Early Notification of a Nuclear Accident – A 1986 Convention that aims to provide relevant information concerning nuclear accidents as early as possible to minimize transboundary radiological consequences.

Convention on International Trade in Endangered Species of Wild Fauna and Flora (CITES) – CITES bans or regulates international trade in endangered species to combat over their exploitation. The text was opened for signature in Washington D.C. on 3 March 1973 and entered into force on July 1, 1975. (CITES)

Convention on Nuclear Safety – The Convention establishes a legal obligation on the part of the Parties to apply certain general safety principles to the construction, operation, and regulation of land-based civilian nuclear power plants under their jurisdiction and submit periodic reports on the steps they are taking to implement the obligations of the Convention. The Convention, which was opened for signatures in September, 1994 and entered into force in 1996.

Convention on Supplementary Compensation for Nuclear Damage – The Convention on Supplementary Compensation for Nuclear Damage was adopted on 12 September 1997 and was opened for signature at Vienna on 29 September 1997 at the 41st General Conference of the International Atomic Energy Agency. The Convention allows for compensation for damage within a country's exclusive economic zone, which includes loss of tourism or fisheries related income.

Convention on the Conservation of Migratory Species of Wild Animals – A 1979 UNEP sponsored Convention and its amendments to protect the habitat of migrating species whether within a single country or of a transboundary nature. (UNEP)

Convention on the Elimination of All Forms of Discrimination Against Women (CEDAW) – CEDAW was adopted in 1979 by the UN General Assembly and is often described as an international bill of rights for women. Consisting of a preamble and 30 articles, it defines what constitutes discrimination against women and sets up an agenda for national action to end such discrimination. (CEDAW)

Convention on the Physical Protection of Nuclear Material – This 1980 Convention provides for certain levels of physical protection during international transport of nuclear material and establishes a general framework for cooperation among states in the protection, recovery, and return of stolen nuclear material.

Convention on the Rights of the Child (CRC) – The treaty which fully articulates the human rights of children and the standards to which all governments must aspire. The Convention, entered into force in 1989, and is administered by UNICEF. Ratified by nearly all countries, it is the most universally accepted human rights instrument in history.

Cooperation – The idea that agents can increase their fitness by mutual help rather than by competition. This assumes that resources adequate for both exist, or are created by the interaction. (CSG)

Coordinating Group of Amazon Basin Indigenous Organizations (COICA) – COICA is a coordinating body for more than four hundred indigenous groups. Now headquartered in Quito, Ecuador, it was founded in Lima Peru in 1982 to help native people defend their rights, fight for the survival of their culture, to exchange experiences and to find solutions to their various common problems. (COICA)

CORINE – A program for the Coordination of Information on the Environment proposed in 1985 by the European Commission that is to gather information on priority environmentally related topics for the European Union. (EEA)

Corridor – A band of vegetation, usually older forest, which serves to connect distinct patches on the landscape.

Corruption – Inducement to wrong by improper or unlawful means such as bribery. (MW)

Cost-benefit analysis – A decision tool that assesses projects through a comparison between their costs and benefits, including social costs and benefits. (DFID)

Cost-benefit approach – A method of analysis used in circumstances where components of the real costs or benefits of a project would not be adequately represented by market prices and may not be traded at all. It assumes that benefits and costs are ultimately defined in terms of human preferences.

Cost-effectiveness – The relation between the costs (inputs) and outputs of a project. The most cost-effective is the most economical in terms of tangible benefits produced by money spent.

Cost-efficiency – The relation between the costs (inputs) and outputs of a project. A project is most cost-efficient when it achieves its outputs at the lowest possible cost compared either to alternative methods of delivery or alternative outputs achieving the same objective. (DFID)

Cotonou Agreement – A partnership agreement between the EU and the countries of Africa, the Caribbean and the Pacific signed in June 2000 in Cotonou, Benin. The agreement replaced the Lomé Convention and has poverty reduction as its main objective, "to be achieved through political dialogue, development aid and closer economic and trade cooperation." (AD)

Council (1) – An executive or governing body whose members are equal in power and authority.

Council (2) – The GEF governing body that consists of 32 members representing constituency groupings that meets semi-annually. It develops, adopts and evaluates the operational policies and programs for GEF-financed activities. (GEF)

Council of Europe – The Council of Europe is an intergovernmental organization established to protect human rights, pluralist democracy and the rule of law; promote awareness and encourage the development of Europe's cultural identity and diversity; seek solutions to problems facing European society; and help consolidate democratic stability in Europe by backing political, legislative and constitutional reform.

Countries with economies in transition (EIT) – The countries of Central and Eastern Europe and former republics of the Soviet Union that are in transition to a market economy.

Country of origin of genetic resources – The country that possesses those genetic resources in *in-situ* conditions (CBD).

Country profile – A socioeconomic, political, environmental and demographic overview of a nation that emphasizes the interests of the institution making the overview.

C

Credentials – A document issued by a State authorizing a delegate or delegation of that state to attend a conference, including, where necessary, for the purpose of negotiating and adopting the text of a treaty. (UNT)

Creditor Reporting System (CRS) – A database jointly established by the OECD and the World Bank in 1967 that reports Official Development Assistance (ODA) and Official Aid (OA) and other lending to developing countries and countries in transition by OECD member countries and Bretton Woods institutions.

Crises management – Management that responds to unforeseen circumstances with no time to plan ahead. (DFID)

Crosscutting issues (also cross-sectoral issues) – Topics of concern to several different sectors or interests that include subjects such as education, finance and budgeting, personnel management and security, trade, technology transfer, consumption and production patterns, science, capacity building and information.

Cultivar – A cultivated variety (genetic strain) of a domesticated crop plant. (GBS)

Cultural diversity – The variety or multiformity of human social structures, belief systems, and strategies for adapting to situations in different parts of the world. (GBS)

Customary international law – Usual and expected, although unwritten, rules of behavior regarding relations between states.

D

DAC – See Development Assistance Committee.

Daily Subsistence Allowance (DSA) - An amount of money given to a traveller on official mission to cover housing and food expenses. (UN)

Darwin Declaration – A declaration by the Darwin Workshop on Removing the Taxonomic Impediment (February 1998) that recognizes the lack of sufficient infrastructure, training, research and access to taxonomic information required for the conservation, sustainable use and equitable sharing of benefits from biodiversity.

Davos (Switzerland) Symposium – See the World Economic Forum (WEF).

De facto – In fact (as opposed to in law, *de jure*). (BLD)

De jure – In law (as opposed to in fact, *de facto*). (BLD)

De lege ferenda – What the law ought to be (as opposed to what the law is. (BLD)

Debt relief – Debt relief may take the form of refinancing, re-organization, rescheduling or cancellation of all repayments. A loan is refinanced when the creditor country makes a new loan to enable the debtor country to meet the debt service payments on an earlier loan. A loan is rescheduled when the amortization or interest payments or both on the outstanding portion are rearranged to make payment easier. (DFID)

Decentralization – The general term for a transfer of authority and/or responsibility for performing a function from the top management of an organization or the central governance level of an institution to lower level units or the private sector. (UNDP)

Decentralized energy path (DE) – The generation of electricity at the point of its use regardless of plant size, fuel or technology.

Decision – A formal agreement that leads to binding actions. It becomes part of the agreed body of decisions that direct the work of a COP or other governing body.

D

Decision 21/21 – A decision of the UNEP Governing Council that established an Open-ended Intergovernmental Group of Ministers to undertake a comprehensive policy-oriented assessment of weaknesses in existing international environmental institutions and examine options for strengthened international environmental governance.

Declaration – A joint statement by two or more states that is binding on those states making the declaration (mandatory declaration). Non-mandatory declarations may be non-binding. The term may also refer to an "interpretive" declaration from a state as to its understanding of some matter covered by a treaty but these are not the same as reservations. (UNT)

Declaration and Plan of Action of Barbados – The formal statement produced by participating states in the Global Conference on the Sustainable Development of Small Island Developing States (SIDS) held in Bridgetown, Barbados, 25 April-6 May 1994. The Barbados Declaration made a commitment to specific policies, actions, and measures to be taken at the national, regional, and international level to enable SIDS to achieve sustainable development. It recognized the special constraints to sustainable development that SIDS face economically, environmentally, and socially by virtue of their small size. It called for the assistance of regional and international entities, both governmental and non-governmental, to engage in a program of action aimed at the sustainable development of the SIDS.

Definitive signature – A way of establishing the consent of the state to be bound by a treaty that is not subject to ratification, acceptance or approval. (UNT)

Deforestation – The change of forested lands to non-forest lands. The removal of forest stands by cutting and burning to provide land for agricultural purposes, residential or industrial building sites, roads, etc., or by harvesting the trees for building materials or fuel. (IPCC)

Delegation – A group representing others, which may include high-level representatives of relevant ministries including foreign affairs, technical experts, and increasingly, NGO representatives; the grant of authority to a person to act on behalf of one or more others, for agreed purposes.

Demarsch – A diplomatic or political initiative or maneuver. A petition or protest presented through diplomatic channels. (MW).

Demographic, Environmental, and Security Issues Project (DESIP) – A project that is to help illuminate connections between environmental scarcity and political conflict. The project was to compile information on the world's wars and trouble spots but now includes information on the demographic, political and environmental factors that impact human and other life on the planet. (DESIP)

Depository – The institution acting as the custodian of a treaty and that is entrusted with several specific functions regarding the commitments of the parties: accepts notifications and documents related to treaties, examines whether all formal requirements are met, registers and deposits all related documents, and notifies the parties of all relevant actions regarding the treaty. (UNT)

Dept for nature swap – An arrangement in which an indebted developing country undertakes, in exchange for cancellation of a portion of its foreign debt, to establish local currency funds to be used to finance a conservation program.

Desertification – The progressive destruction or degradation of vegetative cover especially in arid or semiarid regions bordering existing deserts.

Developed country – A highly industrialized country usually from the North or West.

Developing country – Low and middle-income countries in which most people have a lower standard of living with access to fewer goods and services than do most people in high-income countries. There are currently about 125 developing countries with populations over 1 million; in 1998, their total population was more than 5.0 billion. The term is used to categorize countries eligible for Official Development Assistance. (WB)

Development – Process of increasing capacities in the context of a defined goal or set of goals, commonly of a particular sphere at the level of either a community, state or an organization (social development, economic development, institutional development, etc.) (UNDP) A process made up of activities leading to the use, improvement, or conservation of natural and economic goods and services in order to maintain and improve human life quality of a target population. (OAS) The growth of a person or society toward its full potential. It is often used to mean "industrialization" or "modernization" of countries of the South, with the underlying assumption that the industrial model is worth emulating. (R)

Development Assistance Committee (DAC) – A forum for consultation among most of the donor countries, together with the EC, on how to increase the level and effectiveness of aid flows to all aid recipient countries. DAC sets the definitions and criteria for aid statistics internationally. (DFID) The following includes member states and supporting agencies within those countries.

Australia
- AusAid, Australian Agency for International Development

Austria
- BKA, Federal Chancellery
- BMA, Federal Ministry of Foreign Affairs
- OeKB, *Oesterreichische Kontrollbank* AG

Belgium
- DGIC, Directorate General for International Cooperation
- MF, Ministry of Finance

Canada
- CG, Canadian Government
- CIDA, Canadian International Development Agency
- IRDC, International Development Research Centre

Denmark
- DANCED, Danish Co-operation for Environment and Development
- DANIDA, Danish International Development Agency
- MFA, Ministry of Foreign Affairs

European Community
- EDF, European Development Fund
- EIB, European Investment Bank

Finland

- MFA, Ministry of Foreign Affairs

France

- AFD, *Agence Française Développement*
- FSP, *Fonds de solidarité prioritaire*
- MAE, *Ministére des Affaires Étrangéres*
- Natexis, *Nateis Banque Populaire*

Germany

- BMZ, *Bundesministerium für Wirtschaftliche Zusammenarbeit und Entwicklung*
- FO, Foreign Office
- GTZ, *Duetsche Gesellschaft für Technische Zusammenarbeit*
- KFW, *Kreditanstalt für Wiederaufbau*
- LG, Federal States and Local Governments

Ireland

- DFA, Department of Foreign Affairs

Italy

- CA, Central Administration
- DGCS, *Direzione Generale per la Cooperazione allo Sviluppo*
- LA, Local Administration
- MC, *Mediocrecito Centrale*
- SACE, *Sezione Speciale perl'Assicurazione del Credito all'Esportazione*

Japan

- JBIC, Japan Bank for International Co-operation
- ICA, Japan International Co-operation Agency
- MOFA, Ministry of Foreign Affairs

Netherlands

- MFA, Ministry of Foreign Affairs

Norway

- MFA, Ministry of Foreign Affairs
- NORAD, Norwegian Agency for International Development

D

Portugal
- GP, Portuguese Government
- ICP, Portuguese Co-operation Institute

Spain
- ECON, Ministry of Economy and Finance
- EDUC, Ministry of Education and Science
- ENV, Ministry of Environment
- ICO, *Instituto de Credito Oficial*
- MFA, Ministry of Foreign Affairs

Sweden
- Sida, Swedish International Development Authority

Switzerland
- DDC, *Direction de Développement et de la Coopération*
- SECO, *Secrétariat d'État á l'Économie*

United Kingdon
- DFID, Department for International Development
- CDC, CDC Capital Partners PLC

United States
- AID, U.S. Agency for International Development
- STATE, U.S. State Department
- TDA, U.S. Trade and Development Agency

Development cooperation – A term often used synonymously with overseas development to indicate the interdependent nature of the development process and to emphasize the cooperation between poorer and richer countries.

Development round – WTO negotiations initiated in Doha, Qatar in 2001 that emphasized to eliminate trade protectionism (primarily in agriculture).

Diplomacy – The art and skill of conducting negotiations between states or between a state and an international organization. A form of international dispute settlement that attempts to reconcile parties to a disagreement by use of negotiation, mediation, or inquiry. (MW)

Diplomat (1) – An informal term for government official who is a skilled negotiator for the position of a state. (MW)

Diplomat (2) – A rank entitling a representative of a state to certain privileges such as immunity from arrest and prosecution when on official status or performing official duties. (MW)

Diplomat (3) – A rank reserved for the head of an international organization, generally carrying the same privileges and immunities as a diplomat of a state. (MW)

"Diplomatic License ®" – A CNN weekly program hosted by Richard Roth, which examines issues affecting the international community, the UN and its specialized organizations.

Direct use value – An economic value derived from the direct use or interaction with a resource or resource system. (GBA)

Disaster management – A collective term employed to encompass all aspects of planning for and responding to disasters. (PAHO)

Disaster mitigation – Long-term measures to reduce the scale and/or the duration of adverse effects of hazards on a society or community which is at risk, by reducing the vulnerability to its people, structures, services and economic activities. (OAS)

Disaster prevention – Measures aimed at reducing the chances of occurrence of a hazard event and/or preventing or reducing the impact of the disaster. (DFID)

Division for the Advancement of Women (DAW) – DAW was established in 1946 as the "Section on the Status of Women, Human Rights Division, Department of Social Affairs" of the UN. In 1972, the section was upgraded to the "Branch for the Promotion of Equality for Men and Women" and in 1993, it was moved to New York as a part of the Department of Policy Coordination and Sustainable Development (DPCSD) which, in 1996 became the Department of Economic and Social Affairs (DESA). DAW acted as the substantive secretariat for the Fourth World Conference on Women in Beijing (1995), the largest conference in the history of the United Nations. (DAW)

D-level staff – A term used to categorize director-level/professional level staff (and salary/benefit package) in the UN and OAS systems, starting at D1 and rising to D2. (UNDP)

D

Doha Ministerial Declaration – The final document of the World Trade Organization's fourth ministerial conference (November 2001) in which the 142 members of WTO decided to launch a new three-year round of negotiations (the Doha Development Agenda) based on further trade liberalization and commitments to build capacity in developing countries. It also approved the accession of China and Taiwan to the WTO.

Domestic benefits – The benefits that accrue to a country from a GEF project aimed at achieving global environmental benefits. The country bears the expenditure for the domestic benefits. (GEF)

Donor country – A nation that participates in the direct transfer/contribution of funds, expertise, or equipment through grants or low interest loans to developing countries.

Double majority – A double majority requires a majority to be maintained both on the basis on one vote per country, as well as on the basis of financial contributions to a fund. (JG)

Double weighted majority system – System used in the formal vote procedure, it is an affirmative vote representing both a 60% majority of the total number of participants and 60% majority of the total contributions. (GEF)

Draft agenda – A proposed program of work not yet formally approved by a meeting.

Drafting group (or committee) – Group established by the Chair or President to facilitate negotiations and to prepare text. Normally, observers may not attend meetings of the drafting groups.

Drainage basin – An area of the earth's surface from which surface drainage all flows to a single outlet stream (a watershed in North America).

Drought – A period of deficient precipitation or runoff extending over an indefinite number of days, but with no set standard by which to determine the amount of deficiency needed to constitute a drought. Thus, there is no universally accepted quantitative definition of drought; generally, each investigator establishes his or her own definition. (USGS)

Daily Subsistence Allowance (DSA) - An amount of money given to a traveller on official mission to cover housing and food expenses. (UN)

Dublin Statement – The Dublin Statement on Water and Sustainable Development adopted at the International Conference on Water and the Environment (ICWE), held in Dublin, Ireland, January 1992. The ICWE was attended by over 500 participants from more than 100 countries and 80 international governmental organizations and NGOS.

Dynamics – The behavior of a system in time. Changes with time are the essence of complexity; a static system is merely a snapshot within an evolutionary continuum, however interesting it may be in its own right. (CSG)

E

Earth Charter – An initiative to create an instrument that would set forth fundamental principles for sustainable development called for in 1987 by the United Nations World Commission on Environment and Development and in 1994 by the Earth Council and Green Cross International with support from the Dutch government. An Earth Charter Commission was formed in 1997 to oversee the project and an Earth Charter Secretariat was established at the Earth Council in Costa Rica. (EC)

Earth Council, The – The Earth Council is an international non-governmental organization (NGO) that was created in September 1992 to promote and advance the implementation of the Earth Summit agreements. It is led by a body of 18 Members, drawn from the world's political, business, scientific and non-governmental communities. (EC)

Earth Negotiations Bulletin – A reporting service for environment and development negotiations published by the International Institute for Sustainable Development. (IISD)

Earth Summit – A common name for the 1992 UN Conference on Environment and Development (UNCED) from which came five major documents: the Rio Declaration, *Agenda 21*, A Framework Convention on Climate Change, A Framework Convention on Biological Diversity and a Statement of Principles on Forests.

Earth Summit +5 – See UNGASS.

Earth system – The Earth regarded as a unified system of interacting components, including geosphere (land), atmosphere (air), hydrosphere (water and ice), and biosphere (life). (NASA)

Ecocertification – Process leading to a production unit being awarded a certificate by an independent body, confirming that quality criteria for management of natural resources and other environmental quality criteria have been met. Ecocertification works on a voluntary basis: companies that desire to have a good environmental image are free to take part in this process. For the consumer, the visible part of ecocertification is an "ecolabel" on the finished product, which is recognized as a guarantee of management quality.

Ecodevelopment – Concept developed by Ignacy Sachs (1976) that combined basic needs strategies with an environmental inspired in large part by the 1972 UN Stockholm Conference on the Human Environment. Ecodevelopment is directed at the rural poor and embraces an environmentally conservative doctrine.

Eco-efficiency – The delivery of competitively priced goods and services that satisfy human needs and bring quality of life, while they progressively reduce ecological impacts and resource use intensity throughout the life cycle to a level at least in line with the earth's estimated carrying capacity. Eco-efficiency is based on issues of economic efficiency that will produce environmental benefits, while cleaner production starts from issues from issues of environmental efficiency that will produce economic benefits. (WBCSD)

EcoFund '92 – An independent, non-profit organization for mobilizing private funds to finance non-governmental activities in support of UNCED preparations.

Ecological corridor – See Biological Corridor.

Ecological Economics – A branch of economics that takes account of ecological principles and examines economic values of non-market products and services. (BCHM)

Ecology – The scientific discipline that is concerned with the relationships between organisms and their past, present, and future environments. (ESA)

Eco-Management and Audit Scheme (EMAS) – The result of a European Union directive for a framework to implement and police "green" labeling schemes in EU member states. (DFID)

Economic development – The process of improving the quality of human life through increased per capita income, reduced poverty, enhanced individual economic and educational opportunities, better health and nutrition, resource conservation, and a cleaner environment. (DFID)

E

Economic growth – An increase in a country's total output. It may be measured by the annual rate of increase in a country's Gross National Product (GNP) or Gross Domestic Product (GDP) as adjusted for price changes. (DFID)

Economic valuation – A method of giving economic value to environmental factors and considerations which helps give weight to such considerations where they might otherwise not be taken into account.

Ecosystem – A dynamic complex of plant, animal and microorganism communities and their non-living environment interacting as a functional unit. Ecosystems boundaries are not fixed and their parameters are set according to the scientific, management, or policy question being examined.

Ecosystem approach – For the purposes of the CBD, the "ecosystem approach" is a strategy for the integrated management of land, water, and living resources that promotes conservation and sustainable use of the components and processes of an ecosystem. It recognizes that humans, with their cultural diversity, are an integral component of ecosystems. (CBD)

Ecosystem health – See "Clarity and Chaos" section of the essay.

Ecosystem restoration – See "Clarity and Chaos" section of the essay.

Ecosystem services – If any aspect of ecosystem structure and function has a value, known or unknown, to its inhabitants, it can be classified as an ecosystem service. For example, photosynthesis, a natural ecosystem process, provides food and fiber, and storage of high water by wetlands provides a flood control service. Ecosystem services may have economic, social or cultural values and, therefore, they are important to current human activities; the may have scientific values and are important for future development; and they may control ecosystem functioning and are, therefore, important to a sustained flow of other goods and services. Such services are variously named in the literature as "natural services," "nature"s services" and "natural or system functions." (OAS)

Ecotourism – Travel and travel services undertaken to visit sites or regions of unique natural or historic quality.

Edge of chaos (EOC) – Point between chaos and stasis where evolution is most likely to occur. The tendency of dynamic systems to self-organize to a state roughly midway between globally static (unchanging) and chaotic (random) states. (CSG)

Efficiency – Achievement of a goal at the lowest cost.

El Niño – A climatic event occurring irregularly, but generally every 3 to 5 years so named because it first became evident during the Christmas season at the surface of the eastern tropical Pacific Ocean in part created by seasonal changes in the direction of the tropical winds over the Pacific and abnormally warm surface ocean temperatures. Although most intense in the Pacific region, changes can disrupt weather patterns throughout the tropics and to higher latitudes.

Emergence – System properties that are not evident from those of the parts. Higher-level phenomena that cannot be reduced to that of the simpler constituents and needs new concepts to be introduced. (CSG)

Emission permit – A non-transferable, non-tradable allocation of entitlement by a government to an individual firm to emit a specified amount of a substance. (EM)

Emissions – The release of green house gases and/or their precursors into the atmosphere over a specified area and period of time. (UNFCCC)

Emissions standards – The maximum amount – either rate or concentration – of a particular pollutant that may legally be released into the air from a single pollutant source.

Emissions trading – Emissions trading is a regulatory program that allows firms the flexibility to select cost-effective solutions to achieve established environmental goals. With emissions trading, firms can meet established emissions goals by: (a) reducing emissions from a discrete emissions unit; (b) reducing emissions from another place within the facility; or (c) securing emission reductions from another facility. Emissions trading encourages compliance and financial managers to pursue cost-effective emission reduction strategies and incentivizes emitting entrepreneurs to develop the means by which emissions can inexpensively be reduced. (CF)

Empowerment – The expansion of people's capacities and choices; the ability to exercise choice based on freedom from hunger, want and deprivation; and the opportunity to participate in, or endorse, decision-making that affects their lives. (UNEP) The expansion of freedom of choice and action, generally a participatory process which places or transfers decision-making responsibility and the resources to act into the hands of those who will benefit. This can include (i) capacity building for stakeholder organizations; (ii) strengthening legal status of stakeholder organizations; (iii) stakeholder authority to manage funds, hire and fire workers, supervise work, and procure materials; (iv) stakeholder authority to certify satisfactory completion of project and establish monitoring and evaluation indicators and (v) support for new and spontaneous initiatives by stakeholders. (WB)

Enabling activities – Pre–project activities which include inventories, information gathering, policy analysis, and the development of strategies and actions plans that can provide a basis for making decisions regarding funding through the Global Environment Facility.

Enabling environment – Conditions surrounding an activity or system that facilitate the fulfillment of the potential of that activity or system. A policy document concerned with the preconditions for sustainable human development, including supportive laws and regulations, adequate resources and skills, broad understanding and acceptance of the differing roles of the state, private sector and civil society in sustainable human development, a common purpose and trust. (UNDP)

Endangered – IUCN category of taxa in danger of extinction and whose survival is unlikely if the causal factors continue operating. Included are taxa whose numbers have been reduced to a critical level or whose habitats have been so drastically reduced that they are deemed to be in immediate danger of extinction. (IUCN)

Endemic – Restricted to a specified region or locality. (GBS)

Enhanced Structural Adjustment Facility (ESAF) – A facility to provide medium term balance of payments assistance to countries facing protracted balance of payments difficulties.

Entry into force – Multilateral treaties provide for a fixed number of states to express their consent and/or ratify an agreement, convention or treaty for entry into force. Other conditions are a fixed amount of time elapsed or a given percentage of states falling within a defined category. When these conditions are met a treaty enters into force for those states that gave the required consent. (VC)

Environment – See "Clarity and Chaos" section of the essay.

Environment (Human) – The compendium of natural, social, and cultural values existing in a given place and moment that influences the material and psychological life of man. (PL)

Environment Fund – A voluntary fund of UNEP established to provide additional financing for environmental programmes under the guidance of its Governing Council. (UNEP)

Environmental assessment – A term used almost interchangeably with environmental impact assessment, environmental appraisal, and environmental analysis that refers to a formal procedure structured to ensure that selected environmental issues are considered in the early stages of the project cycle.

Environmental auditing – A process of assessing the impacts of existing developments, policies and projects, as opposed to environmental assessments which apply to new or greatly modified developments.

Environmental governance – Process that links and harmonizes policies, institutions, procedures, tools, and information to allow participants (public and private sector, NGOs, local communities) to manage conflicts, seek points of consensus, make fundamental decisions, and be accountable for their actions. (IDB)

Environmental impact assessment (EIA) – A process of evaluating and suggesting management and mitigation scenarios for the impacts arising for a new development at the various stages of the project cycle.

Environmental Integrity Group (EIG) – A coalition working within the UNFCCC that includes Mexico, the Republic of Korea and Switzerland.

Environmental management – The mobilization of resources and the use of government to control the use, improvement or conservation of both natural and economic goods and services in such a way that conflicts created by that use, improvement or conservation are minimized. (OAS)

Environmental Management Group (EMG) – An institution established by the UN General Assembly that focuses on the linkages between environment and development. Chaired by the Executive Director of UNEP, the goal of the EMG is to coordinate and promote joint action within the United Nations in the areas of environmental management and human settlements. (UNEP)

Environmental protection – Actions at the international, national and local levels to prevent and ameliorate deterioration of human environments. Such actions include conservation efforts, recycling, waste reduction and disposal, and the development of cleaner and safer technologies.

Environmental quality – The relative capacity of an environment to satisfy the needs and wants of the individual or society "surrounded" by that environment. (OAS)

Epistemic community – As defined by Haas (1992), an epistemic community is a "network of professionals with recognized expertise and competence in a particular domain and an authoritative claim to policy-relevant knowledge within that domain or issue area. They can be identified by a shared set of normative and principled beliefs, shared causal beliefs, shared notions of validity, and a common policy enterprise."

Equal per capita concept – A concept that implies that the total national "right to emit" carbon is based on population size and a basic per capita emissions quota.

Equity – Impartial or just treatment, requiring that similar cases be treated in similar ways.

Equity Principle – A principle relating to burden sharing and the distribution of costs and benefits that can be temporal (intergenerational) or spatial (among countries and communities).

Erga omnes – "Toward all," wrongful acts that harm everyone and not simply one injured party. (BLD)

Ethical values – Statements of ethical principle that inform the private and social valuation of biological resources. (GBA)

European Bank for Reconstruction and Development (EBRD) – The European Bank for Reconstruction and Development – a major multilateral donor located in Luxembourg.

European Development Fund – A development fund committed to under the Lomé Convention.

European Investment Bank (EIB) – A financial institution of the European Community which provides long term lending to the member states of the European Union approximately 50 percent of which is to the private sector.

European Union – An economic and political alliance currently comprised of 25 European nations founded in 1993. Members cooperate on a number of areas, ranging from a single free trade market to foreign policy. It has a number of institutions including the Council of the EU, the European Parliament, European Commission for civil service, Court of Justice and the Court of Auditors. Preceded by the EEC and EC.

Eutrophication – The nutrient enrichment of waters which results in the stimulation of an array of symptomatic changes, among which increased production of algae and macrophytes, deterioration of water quality and other symptomatic changes, are found to be undesirable and interfere with water uses. (UNESCO) A process by which a body of water becomes enriched in dissolved nutrients (as phosphates) that stimulate the growth of aquatic plant life usually resulting in the depletion of dissolved oxygen. (MW)

Ex propio motu – On its own accord. (BLD)

Exalted rank – Refers to a non-elected person who receives his/her title from blood lines (inherited) or marriage (appointed). (MW)

Excellency – A title generally reserved for a Head of State or Cabinet-level Minister. (MW)

E

Exclusive economic zone (EEZ) – Exclusive Economic Zone. Within this zone (which may extend to 200 nautical miles under UNCLOS III), the state has jurisdiction and control over the exploration, exploitation, management, and conservation of the natural resources of the waters, seabed, and subsoil. Ships and aircraft enjoy high seas freedoms of navigation and over flight unless they infringe upon the coastal state's economic rights within the EEZ.

Executing agency – Within the GEF, the organization that actually carries out the project work in the field with a team of national counterparts. (GEF)

Existence value – The value of knowing that a particular species, habitat or ecosystem does and will continue to exist. It is based on the perceived value of the environmental asset unrelated to current or future use.

Expert group – A group of academics, scientists, government representatives from specialized ministries or departments and NGOs convened to address specific issues without regard to political interest.

Ex-post evaluation – An evaluation of a completed project carried out either immediately after project completion or at a later time when the project impacts are fully developed.

***Ex-situ* collection** – A collection of plant genetic resources for agriculture that are maintained outside their natural habitat (CBD).

***Ex-situ* conservation** – The conservation of plant genetic resources for food and agriculture outside their natural habitats (CBD).

External forcing – Influence on the Earth system (or one of its components) by an external agent such as solar radiation or the impact of extraterrestrial bodies such as meteorites. (NASA)

Externality – An external effect; interactions between economic agents not involving market transactions. By way of their production or consumption activities, economic agents may unintentionally generate positive or negative effects for the well-being of other agents. Externalities may be negative: an industry whose waste pollutes a river might affect the welfare of fishermen. Conversely, conservation of local plant varieties by farmers, who are not remunerated for this, is a positive externality. The costs (or benefits) imposed on one individual or firm by the actions of another, for which the latter does not pay (or receive) compensation.

Extinct – IUCN category *Ex* for taxa that are no longer known to exist in the wild after repeated searches of their type localities and other known or likely places.

Extractive industries review (EIR) – A process that the World Bank Group has launched to discuss its future role in the extractive industries with concerned stakeholders. The aim is to produce a set of recommendations that will guide involvement of the World Bank Group in the sectors of oil, gas and mining. (WB)

Extrabudgetary – Funding outside of the core or regular funds of an agency or organization. (IDB)

Extreme poverty – A situation in which a person or household lacks the resources to consume a certain minimum amount of food judged to be necessary for adequate nutrition, even in the case when all resources are devoted to food.

Extrinsic value – A form of judgment that allows a continuum of possibilities, i.e. a measurement of goodness or presence. (CSG)

F

Factor 4 / Factor 10 – The concept that, in order for sustainability to be reached and maintained during a period when human populations are likely to double in size and average living standards to increase significantly, industry should increase its resource conversion efficiency by a minimum factor of 4 (i.e. resource productivity should grow fourfold). Put another way, the amount of wealth extracted from one natural resource unit should be quadrupled. Since industrialized societies typically consume 20 to 30 times more than less developed ones, some people are calling for a factor 10 increase in conversion efficiency in the developed world. (Rocky Mountain Institute and Wuppertal Institute)

Fair trade – 1. Any exchange of goods as long as it does not involve: forbidden commodities, usury, uncertainty, fraud, or extortion. (Shari'ah). 2. A system of international commerce based on equitable relationships between producers and consumers. The goal is for producers, often farmers in impoverished areas, to receive more of what consumers pay for the goods. Specifically, a fair-trade organization pays producers more than the market rate, provides credit to the producers, builds long-term relationships with them, encourages them to form democratic co-ops, encourages ecologically sustainable production and bypasses intermediaries between producers and consumers. (RCA)

FAO – See **United Nations Food and Agriculture Organization**.

FAO Code of Conduct for Responsible Fisheries (1995) – This Code sets out principles and international standards of behavior for responsible practices with a view to ensuring the effective conservation, management and development of living aquatic resources, with due respect for the ecosystem and biodiversity. The Code recognizes the nutritional, economic, social, environmental and cultural importance of fisheries and the interests of all those concerned with the fishery sector. The Code takes into account the biological characteristics of the resources and their environment and the interests of consumers and other users.

Fast track authority – In the United States it is the authority given to the President by the Congress that allows the President to send to the Congress completed international trade agreements for a vote of "yes" or "no." Under fast track authority, Congress is not be allowed to alter or amend the agreements.

Fast tracking – The process of reducing the number of sequential relationships and replacing them typically with parallel relationships, usually to achieve shorter overall durations but often with increased risk.

Fauna – All the animal species found in a given area.

Feedback – A linking of the output of a system back to the input. Traditionally this can be negative, tending to return the system to a wanted state, or positive tending to diverge from that state. (CSG)

Fertility rate – The average number of children that would be born alive to a woman during her lifetime if she were to bear children at each age in accord with prevailing age-specific rates. (UNCHS)

Final Act of the Uruguay Round for Establishing the World Trade Organization – A report that includes a section on "Decision on Trade and Services and the Environment" which outlined the need to negotiate issues of trade and their relationship with environment with a sustainable development context. These issues are being discussed at the regional and sub-regional levels in a variety of fora as a means of informing the WTO. The Uruguay Round Agreement also formally established the Trade and Environment Committee.

Financial mechanism – A mechanism of the UNFCCC which is to enable the transfer of funds and technology to developing countries on a grant or concessional basis, under the guidance of the COP.

First World – A term refering to developed (in general OECD) countries.

Flexibility mechanisms – A term which refers to the three cooperative implementation mechanisms under the Kyoto Protocol (Joint Implementation, International Emissions Trading and Clean Development Mechanism) and which includes the notion of differentiated commitments. (EM)

Flora – All the plant species found in a given area.

Force majeure – A superior or irresistible force; an event or effect that cannot be reasonably anticipated or controlled. (MW)

Fluorocarbons – Carbon-fluorine compounds that contain other elements such as hydrogen, chlorine, or bromine. Common fluorocarbons include chlorofluorocarbons (CFCs), hydrochlorofluorocarbons (HCFCs), hydrofluorocarbons (HFCs), and perfluorocarbons (PFCs).

Focal area (also thematic area) – The four main areas that the GEF works in. They are climate change, biological diversity, international waters and ozone layer depletion. (GEF)

Focal point (1) – An arrangement whereby a national entity reflects an international public interest.

Focal point (2) – A formal representative nominated by a government to receive and distribute communications on a given topic as well as to participate in meetings on the subject on its behalf.

Foreign direct investment (FDI) – The act of building productive capacity directly in a foreign country. Investment in one country by firms owned in another country.

Forest landscape restoration (FLR) – A planned process implemented to restore a specific forest type and enhance human well-being in deforested or degraded forest landscapes that requires choices at a landscape scale were overall landscape benefits are more important than choices relating to individual forest stands or sites.

Forest Principles – The non-legally binding authoritative statement of principles to guide the management, conservation and sustainable development of all types of forests agreed to by nearly 180 governments at UNCED in 1992.

Forum – A medium of open discussion or expression of ideas. (MW)

Fossil fuel – Any hydrocarbon deposit that can be burned for heat or power, such as petroleum, coal, and natural gas (or any fuel derived from them). (NASA)

Founex Report – The written product of a meeting in Founex, Switzerland called to prepare for the 1972 Stockholm Conference that discusses the links between development and environmental protection in the Third World. (CIPA)

Fragile ecosystem – (See "Clarity and Chaos" section of the essay).

Fragmentation – The breaking up of extensive landscape features into disjunct, isolated, or semi-isolated patches as a result of land-use changes. (BCHM)

Framework – A broad overview, outline or skeleton within which details can be added.

Framework convention – Agreements under which protocols can be developed to address specific subjects that require more detailed and specialized negotiations. (UNFCCC)

Free trade – Trade in which goods can be imported and exported without barriers in the forms of tariffs, quotas, or other restrictions.

Free-rider problem – In debate of the UNFCCC the "free-rider problem" refers to the possibility that some Parties may take advantage from climate action by others without themselves accepting commitments or taking action.

Friends of the Chair (President) – A few prominent negotiators that are invited by the Chair/President to help informally in developing consensus on issues. (UNFCCC)

Friends of the Earth International (FoEI) – FoEI was founded in 1971 by four organizations from France, Sweden, England and the USA. Today's federation of 66 groups grew from annual meetings of environmentalists from different countries who agreed to campaign together on certain crucial issues, such as nuclear energy and whaling. In 1998, the combined number of members and supporters of Friends of the Earth groups was close to one million, and the FoEI umbrella united almost 5,000 local activists groups. The combined annual budget of FOE groups was close to US$200 million, and together they employed close to 700 full time staff members. (FOE)

Fugitive fuels – Fuel emissions that may or may not be fully controlled although in most cases they are not accidental. Examples of fugitive emissions are leaks from gas pipelines and valves, venting and flaring of gases, methane emissions from coal seams and vapor given off by petroleum stores.

Full powers – A document emanating from the competent authority of a State designating a person or persons to represent the State for negotiating, adopting or authenticating the text of a treaty, for expressing the consent of the State to be bound by a treaty, or for accomplishing any other act with respect to a treaty. (VC)

G

G-7 – Group of seven industrialized democracies (United States, Japan, Germany, France, Britain, Italy, and Canada).

G-8 – Group of eight main industrialized democracies (G-7) and Russia.

G-15 – Group of 15. A summit level group of developing countries organized in 1989 that provides a platform to promote South-South cooperation and North-South dialogue. The group now has twenty members (Algeria, Argentina, Brazil, Chile, Colombia, Egypt, India, Indonesia, Iran, Jamaica, Kenya, Malaysia, Mexico, Nigeria, Peru, Senegal, Sri Lanka, Venezuela, and Zimbabwe).

G-20 – An international forum of finance ministers and central bank governors representing nineteen governments, the European Union and the Bretton Woods Institutions (World Bank and the International Monetary Fund).

G-21 – A group of developing countries that coalesced at the WTO round in Cancun, Mexico in mid-2003 and is dedicated to negotiating an elimination of subsidies in all sectors, and particularly those provided in developed nations to sectors (such as agriculture) which directly impact the ability of developing nations to compete in the international marketplace.

G-77 + China – A coalition of nearly 150 members which aims to collectively articulate and promote the economic interests of the group and to enhance its negotiating position on all major issues in the UN. The chair rotates annually between nations and regions. The coalition currently has a Rome Chapter (at FAO), Paris Chapter (at UNESCO), Nairobi Chapter (at UNEP), Washington Chapter (at IMF and World Bank) and a Vienna Chapter (at UNIDO).

***Gaia* hypothesis** – The hypothesis that the Earth's atmosphere, biosphere, and its living organisms behave as a single system striving to maintain a stability that is conductive to the existence of life. (NASA)

Gap analysis – Systematic application of goals and criteria to determine what resources or values are currently protected and what needs to be protected.

GEF – See Global Environment Facility.

Gender – The socially determined roles played by women and men, which can de determined by historical, religious, economic, cultural, and ethnic factors. (UNDP)

Gene – The functional unit of heredity: the part of the DNA molecule that encodes a single enzyme or structural protein unit. GBS)

General Agreement on Tariffs and Trade (GATT) – Originally signed by 23 nations in 1947, GATT in an international body devoted to eliminating import quotas, lowering tariffs, and otherwise promoting free, nondiscriminatory international trade. The Uruguay Round created the World Trade Organization as an international body in 1995 to administer the GATT.

General Assembly – The highest principle organ of an intergovernmental organization that consists of representatives of all its member states. (UNT)

General circulation model (GCM) – A global, three-dimensional computer model of the climate system which can be used to simulate human-induced climate change. GCMs are highly complex and they represent the effects of such factors as reflective and absorptive properties of atmospheric water vapor, greenhouse gas concentrations, clouds, annual and daily solar heating, ocean temperatures and ice boundaries. The most recent GCMs include global representations of the atmosphere, oceans, and land surface. (NASA)

Genetic diversity – Variation in the genetic composition of individuals within or among species. (GBS)

Genetic material – Any material of plant origin containing functional units of heredity. (CBD)

Genetic resources – Genetic material of actual or potential value. (CBD)

Genetically modified organism (GMO) – An organism whose genetic characteristics have been changed by inserting a modified gene or a gene from another variety or species. Genetically modified organisms may be plants, animals or microorganisms.

Geographic information system (GIS) – A computer based system that stores, manipulates and displays spatial information.

Global benefits – Environmental benefits accrued at the global level as opposed to those gained nationally or locally through the development efforts of a government. (GEF)

Global Biodiversity Forum (GBF) – Mechanism founded in 1993 by IUCN, WRI, UNEP, that acts to encourage analysis, dialogue and partnership on the main ecological, economic, social and institutional issues related to biodiversity at the local, national, regional and international levels.

Global civil society – A term implying an extension of civil society into the transnational realm, where it constitutes something along the lines of a "regime" composed of local, national, and global NGOs. (L)

Global Climate Coalition (GCC) – A coalition of energy based industries that oppose the UNFCCC. (GCC)

Global Compact (GC) – A network of five UN agencies (HCHR, ILO, KNOP, UNIDO, UNEP) with private sector companies and other organizations formed in 2000 and promoting acceptance of 10 principles in the broad areas of human rights, labor, environment, and corruption and aimed at resonsible corporate citizenship

Global commons – A component of the earth-atmosphere system that is claimed by no one but has the potential to be used by everyone – which carries the potential that, over time, it would be overused and destroyed.

Global Development Learning Network (GDLN) – The GDLN is a worldwide partnership of distance learning centers and other public, private and non-governmental organizations committed to development learning and development dialogue for lasting poverty reduction. Offering a unique combination of distance learning technologies and methods, GDLN facilitates timely and cost-effective knowledge sharing, consultation, coordination and training. (World Bank, UNESCO-IHE)

Global dimming – A term which refers to the reduction in the amount of sunlight reaching the earth surface due to increased pollution and water vapor.

Global Environment Facility (GEF) – An intergovernmental institution supported by its members that offers grants to cover the additional costs to a country's own actions for sustainable development to maximize agreed global environmental benefits in the areas of biodiversity, climate change, international waters, ozone layer depleting substances, persistent organic pollutants, and land degradation, primarily desertification and deforestation. Its secretariat is housed in the World Bank and its primary implementing agencies are the World Bank, UNDP, and UNEP. New Executing Agencies under expanded opportunities include UNIDO, UNFAO, the Regional Development Banks and IFAD. These agencies have direct access to GEF Project Development Funds. (GEF)

- GEF Mid-size Project – A GEF grant for one time of up to US$1,000,000.
- GEF Project – A GEF grant that has no set limit.
- GEF Trust Fund (GEFTF) – Fund set up with the World Bank as Trustee, which consists of contributions received from members of the GEF and is used to fund grants. (GEF)
- GEF Working Paper Series – A series of documents that provide general information on GEF's work and more specific information on methodological approaches; scientific and technical issues; and policy and strategic matters. (GEF)
- GEFable – A term used to describe a project proposal that meets the terms and conditions for a GEF grant.
- GEF-PDF "Block A" – A GEF grant of up to US$25,000.
- GEF-PDF "Block B" – A GEF grant of up to US$750,000.
- GEF-PDF "Block C" – A GEF grant of up to US$1,000,000.

Global environmental benefits – Benefits that accrue to the global community, e.g. the reduction of greenhouse gas emissions. (GEF)

Global Mechanism (GM) – Established under the Convention on Desertification, the GM is in charge of promoting actions for the mobilization and channeling of substantial financial resources, including for the transfer of technology, on a grant basis, and/or on concessional or other terms, to affected developing country Parties. (CD)

Global Ministerial Environment Forum – According to GA Resolution 53/242 of 28 July 1999 it is an annual, ministerial level, global environmental forum that meets together with the UNEP Governing Council and, in alternate years, takes the form of a special session of the Governing Council. Participants review important and emerging policy issues with due consideration for the need to ensure the effective and efficient functioning of the governance mechanism of the United Nations Environment Programme, as well as possible financial implications, and the need to maintain the role of the Commission on Sustainable Development as the main forum for high-level policy debate on sustainable development.

Global warming – A theory relating to an increase in the earth's temperature caused, in part, by the "green-house" effect being created by emissions of GHGs associated with human activities.

Globalization (1) – The increasing worldwide integration of markets for goods, services and capital, that attracted special attention in the late 1990s.

Globalization (2) – A term used to encompass a variety of other changes that were perceived to occur at about the same time, such as an increased role for large corporations (MNCs) in the world economy and increased intervention into domestic policies and affairs by international institutions such as the IMF, WTO, and World Bank. (AD)

Globalization (3) – A term used to describe the increasing flow of goods, services, capital, technology, information, ideas, and labour at the global level, driven by liberalization policies and technological change considered by proponents to be an integrating and inclusive force. (UN)

Goal – A set general purpose and direction the end result of ultimate accomplishment toward which an effort is directed.

Good neighborliness principle – A principle of international law which calls for a broad standard of respect and recognition and which obligates states, enterprises and individuals to respect the rights of others to participate in the social development process.

"GOUTHE" of water - Global Observatory of Units for Teaching, Training and Ethics. A UNESCO-IHP initiative aimed at stimulating and structuring cooperation and networking on water-related issues. It will act as a platform and alliance for information and communication, extending geographically limited partners and facilitating new partnerships.

Governance – Concept describing the way power is exercised in the management of a country's economic and social resources through application of responsibility, participation, information availability, transparency, and the rule of law. Governance is not equal to government, which is the art of administration at a given level of power. Rather, it is the art of coordinating administration actions between different territorial levels – one of which may be global. "Governance is the sum of the many ways individuals and institutions, public and private, manage their common affairs." (CGG)

G

Governing Council, UNEP (GC) – The 58-member governing body of the United Nations Environment Programme established in December of 1975. The UNEP/GC reports to the General Assembly through the Economic and Social Council. It responsibilities and functions include:

- promotion of international cooperation in environmental matters,
- general policy guidance for the direction and coordination of environmental programmes within the UN system;
- reviews of global environmental conditions to ensure that emerging problems of international significance receive appropriate and adequate consideration by governments;
- promotion of participation of relevant international scientific and other professional communities for the acquisition, assessment and exchange of knowledge and information, and technical aspects environmental programmes within the United Nations system;
- continuous review of national and international environmental policies and their impact on developing countries including the additional costs that they may incur in the implementation of environmental programmes and projects, and that ensures that such programmes and projects are compatible with the development plans and priorities of those countries; and,
- review and approval of use of the Environment Fund. (UNEP)

Grassroots – People or society at a local level rather than at the center of major political activity.

Gray water/issues – See **Water**.

Greater Mekong Subregion (GMSR) – The name referring to the Mekong River watershed including countries which are members of the Mekong River Commission (Cambodia, Laos, Thailand, Vietnam) and countries which are not members of the Commission (China, Myanmar).

Green productivity – A strategy for enhancing productivity and environmental performance with a view to overall socio-economic development, involving the use of appropriate techniques, technologies and management systems to produce environmentally compatible goods and services. Green productivity can be applied to manufacturing, services, agriculture and the community. (APO)

Green products – Products that have been certified through a process of "eco" or "green" labeling which insures that the product was handled, made or grown under conditions that meet standards of sustained use, pesticide application, and harvesting as well as certain social and economic criteria for workers.

Green sector – A catchall label for "environmental" projects that cover issues related to forests, land, biodiversity and protected areas.

Green water/issues – See **Water**.

Greenhouse gases (GHG) – The gaseous constituents of the atmosphere (water vapor, carbon dioxide, methane, nitrous oxide, and hydrofluorocarbons, perfluocarbons, and sulphur hexafluoride) both natural and anthropogenic, that absorb and re-emit infrared radiation. (UNFCCC)

Greenpeace International – Established in 1971, Greenpeace is an independent campaigning organization that uses non-violent, creative confrontation to expose global environmental problems and to force solutions that are essential to a green and peaceful future. (GP)

GRID – UNEP's Global Resource Information Database (GRID) is a network of information centers that provides improved access to high-quality environmental information. There are currently sixteen GRID centers operational worldwide.

Gross domestic product (GDP) – The measure of output produced by the factors of production located in a country regardless of who owns these factors.

Gross national product (GNP) – The measure of total income earned by the citizens of a country (GNP equals GDP plus earnings from abroad).

GS-level staff – A term used to categorize administrative support staff (and salary/benefit package) in the UN and OAS systems, starting at GS1 and rising to GS7.

Guideline – A recommended or customary method of working to accomplish an objective. A guideline is not enforced but is generally followed.

H

Habitat – The place where an organism lives and/or the conditions of that environment including the soil, vegetation, water, and food.

HABITAT – The United Nations Centre for Human Settlements (UNCHS) was established in October 1978 as the lead agency within the United Nations system for coordinating activities in the field of human settlements. It is the focal point for the implementation of the Habitat Agenda– the global plan of action adopted by the international community at the Habitat II Conference in Istanbul, Turkey, in June 1996. The mission of UNCHS is to promote socially and environmentally sustainable human settlements development and the achievement of adequate shelter for all. (UNCHS)

Habitat Agenda – The second United Nations Conference on Human Settlements, held in Istanbul, June 1996, also called the "City Summit." (UNCHS)

Hard law – The body of binding and enforceable laws (precise, legally binding obligations and appropriate third party delegation) cf. Soft law.

Harmonization – The process of making procedures or measures applied by different countries – especially those affecting international trade – more compatible, as by effecting simultaneous tariff cuts applied by different countries so as to make their tariff structures more uniform.

Heat island effect – A "dome" of elevated temperatures over an urban area caused by the heat absorbed by structures and pavement.

Heavily indebted poor countries (HIPC) – The name given to those poor countries with large debts, the target of initiatives to forgive that debt as a means of assisting development. (AD)

HFC – Hydrofluorocarbon.

High seas – The ocean areas outside the territorial seas and maritime zones of coastal states. They are open to use by all states for peaceful purposes.

High-income country – A country having an annual gross national product

(GNP) per capita equivalent to $9,361 or greater in 1998. Most high-income countries have an industrial economy. There are currently about 28 high-income countries in the world with populations of one million people or more. Their combined population is about 0.9 billion, less than one-sixth of the world's population. (WB)

His/Her Royal Highness – A title generally reserved for a person of "exalted rank" such as a king, queen, prince, or princess. (MW)

Host country – The nation where an office or convention secretariat is located.

Hot air – The quantity of tradable greenhouse gas emissions of a country equal to the difference between its actual emissions and its higher legally binding limit.

Hotspot – An area particularly rich in endemic species that are endangered by land-use change.

Human development index (HDI) – The HDI measures the overall achievements in a country in three basic dimensions of human development: longevity, knowledge and decent standard of living. It is measured by life expectancy, educational attainment (adult literacy and combined primary, secondary and tertiary enrolment) and adjusted income. (ACCU; UNDP)

Human immunodeficiency virus (HIV) – A virus that steadily weakens the body's defense (immune) system until it can no longer fight off infections such as pneumonia, diarrhea, tumors and other illnesses. All of which can be part of AIDS (Acquired Immunodeficiency Syndrome). Most of all HIV infections have been transmitted through unprotected sexual intercourse with someone who is already infected with HIV, by infected blood or blood products (as in blood transfusions), by the sharing of contaminated needles, and from an infected woman to her baby before birth, during delivery, or through breast-feeding. HIV is not transmitted through normal, day-to-day contact. (UN)

Human poverty index – A measure that reflects the distribution of progress and the backlog of deprivations that still exists. (UNDP)

Hydrology – The science dealing with water and snow, including their properties and distribution.

I

ILO Declaration on Fundamental Principles and Rights at Work – A declaration of the 80th session of the ILO in Geneva (June 1998) that all its Members, "even if they have not ratified the Conventions in question, have an obligation arising from the very fact of membership in the Organization to respect, to promote and to realize, in good faith and in accordance with the Constitution, the principles concerning the fundamental rights which are the subject of those Conventions, namely:

- freedom of association and the effective recognition of the right to collective bargaining;
- the elimination of all forms of forced or compulsory labor;
- the effective abolition of child labor; and
- the elimination of discrimination in respect of employment and occupation.

IMO Conventions in the Field of Marine Safety and Prevention of Marine Pollution – A suite of some 33 conventions dealing with marine safety, marine pollution, and liability and compensation, and coordinated through a variety of institutional mechanisms and partnerships by the IMO.
Impermissibility Principle – A principle from the Vienna Convention on Treaties which states that a reservation incompatible with the object and purpose of a convention signed and ratified by the country wishing to make the reservation shall not be permitted.

Implementing agency (IA) – Within the GEF, an IA is one of three agencies (UNDP, UNEP, WB) accountable to the GEF council for activities financed by the GEF. (GEF)

Incremental costs – The additional cost that the GEF funds between the cost of an alternative project that a country would have implemented in the absence of global environmental concerns and a project undertaken with global objectives in mind. (GEF)

Independent state – A state that is sovereign; one that operates independently internationally.

Indeterminate – An IUCN category of taxa known to be Extinct,

Endangered, Vulnerable, or Rare but there is not enough information to say which of the four categories is appropriate.

Indicator – A qualified/quantified parameter that details the extent to which a project objective has been achieved within a given timeframe and in a specified location. (UNDP)

Indigenous peoples development plan (IPDP) – A plan prepared for any investment project which affects indigenous peoples that is designed in a culturally appropriate manner and is based on the full consideration of the options preferred by the indigenous people affected by the project. The Plan also includes provisions which ensure that institutions responsible for government interaction with indigenous peoples should possess the social, technical and legal skills needed to carry out proposed development activities. Elements of an IPDP include: an assessment of the legal framework; collection of baseline data; examination of land tenure; strategy for local participation; design of mitigation measures and activities; assessment of institutional capacity; an implementation schedule and a system for monitoring and evaluation. (WB)

Informal agreement – A common form of international cooperation that permit states to conclude mutually profitable bargains without the formality of treaties.

Informal consultation – An informal meeting not contemplated in the agenda that is called- by the conference Chair who upon receipt of the report of the consultation presents it to the plenary.

Informal contact group – On the instructions of the meeting's president or chair, delegates may meet in private to discuss specific matters in order to consolidate different views, reach a compromise, and produce an agreed proposal, often in the form of a written text.

Informal employment – The percentage of the employed population, men and women, whose activity is part of the informal sector. (UNCHS)

Informal group – A group called into being by one or more of the Parties, for purposes of informal consultation to reach a compromise or to produce an agreed upon proposal. (UNFCCC)

Informal sector – Persons engaged in the production of goods or services with the primary objective of generating employment and incomes to the persons concerned. It includes unregistered commercial enterprises and all non-commercial enterprises that have no formal structure in terms of organization and operation. (UNCHS)

Informal-informal consultation – An informal consultation of parties and at times, observers, to discuss a specific point. The results of the consultancy may then be then presented by a Party in plenary, but without a formal request from the conference chair.

Information and communication technologies (ICTs) – Any equipment, or interconnected system(s) or subsystem(s) of equipment, that is used in the automatic acquisition, storage, manipulation, management, movement, control, display, switching, interchange, transmission, or reception of data or information.

Initiative – An intervention by a formal or informal group to meet a specific objective.

Innovative Learning (I-Learning) - An intelligent blend of (i) technology-based systems, simulation applications and other e-learning packages aimed at bridging distances through internet, video and teleconferencing technologies, and (ii) social-psychological methods that employ tacit and intuitive knowledge and the creative powers of professionals. (Ramsundersingh, UNESCO-IHE)

In-kind contribution – A term generally used to indicate a non-cash contribution to a project or program that may include office space, utility costs, personnel time, equipment, access, etc.

In-material-breach – A term used to indicate a demonstrable non-compliance by a member state with an approved resolution of the UN Security Council or the UN General Assembly.

***In-situ* conservation** – The conservation of ecosystems and natural habitats and the maintenance and recovery of viable populations of species in their surroundings and, in the case of domesticated or cultivated plant species, in the surroundings where they have developed their distinctive properties. (CBD)

Institute@WSIS E-Strategies – An initiative of the Smithsonian Institute and UNDP to promote sustainable development and launched at the WSSD in Johannesburg. It consists of a network of volunteers who deliver short courses in which practical skills and appropriate technologies are transferred directly to participants. (UNDP)

Institutional development – The process of enabling and facilitating an organization to develop and increase its capacities and capabilities so as to meet its desired objectives. (UNDP)

Instrument – A term used to refer to the tools used for certain types of work, which may be physical such as those used in surveying, or notional such as used in establishing law. (UNT)

Integrated area management (IAM) – A management approach whereby a specific area is zoned and regulated for a variety of uses, including research, species protection, tourism, harvesting, cutting of trees, hunting, or fishing that are compatible with the management goals for the area. (BCHM)

Integrated conservation development projects (ICDP) – An approach to biodiversity conservation that reconciles the management of protected areas with the social and economic needs of the allied or nearby local populations. This is usually done by the involvement of these communities in management of the protected areas and by ensuring that their social and economic needs are considered through the development of economic activities. (GEF)

Integrated development planning – A three-part technique for bringing about beneficial social and economic change: (1) diagnosis of principal problems and potentials; (2) preparation of a development strategy; and (3) formulation of a coordinated package of infrastructure, production, and service projects (including conservation) within an action plan for implementing the strategy. (OAS)

Integrated pest management (IPM) – A management system that uses all suitable techniques and methods in as compatible a manner as possible to maintain pest populations at levels below those causing economic injury. (W)

Integrated water resource management (IWRM) – A process which promotes the coordinated development and management of water, land, and related resources in order to maximize the resultant economic and social welfare in an equitable manner without compromising the sustainability of vital ecosystems. (GWP-TAC)

Integration – The action in which smaller and different units are brought together into a larger unit to form a satisfactory and working whole.

Intellectual property right – A right enabling an inventor to exclude imitators from the market for a limited time and generally related to national or international patent laws. (BCHM)

Inter alia – Among other things.

Inter-American Biodiversity Information Network (IABIN) – An initiative of the Summit of the Americas to provide greater and more useful access to biodiversity information throughout the western hemisphere.

Inter-American Development Bank (IDB) – The IDB was established in December of 1959 to help accelerate economic and social development in Latin America and the Caribbean. Its principal functions are to utilize its own capital, funds raised in financial markets, and other available resources, for financing the development of the borrowing member countries; to supplement private investment when private capital is not available on reasonable terms and conditions; and to provide technical assistance for the preparation, financing, and implementation of development plans and projects. Current lending priorities include poverty reduction and social equity, modernization and integration, and the environment. (IDB)

Inter-American Water Resource Network (IWRN) – A network founded in 1993 whose purpose is to strengthen water resource partnerships among nations, organizations and individuals. (OAS)

Intergenerational equity – A core proposition of some definitions of sustainable development which suggests that future generations have a right to an inheritance (capital bequest) sufficient to allow them to generate a level of well-being no less than that of the current generations. (GBA)

Intergovernmental Negotiation Committee (ING) – A formal group of representatives from different interested governments that has been established to negotiate a draft convention or treaty.

Intergovernmental Panel on Climate Change (IPCC) – A group of approximately 2000 scientists established by the World Meteorological Organization and UN Environment Programme (UNEP) in 1988 to assess the state of knowledge on climate change. The IPCC prepares periodic assessments of the scientific basis for understanding global climate change and its consequences. It currently has three Working Groups and two Task Forces: Working Group I assesses the scientific aspects of the climate system and climate change; Working Group II addresses the vulnerability of socio-economic and natural systems to climate change, negative and positive consequences of climate change, and options for adapting to it; Working Group III assesses options for limiting greenhouse gas emissions and otherwise mitigating climate change; the Task Force on National Greenhouse Gas Inventories oversees the National Greenhouse Gas Inventories Programme; and, the Task Group on Scenarios for Climate Impact Assessment. Although the IPPC advises the COP/UNFCCC, it is not an institution of the Convention. (IPPC)

Internal rate of return (IRR) – A decision-making criterion which assesses the amount and timing of project benefits and which is based on a discount rate that reduces the net present value of a project to zero. If the IRR is larger than the discount rate, the action is acceptable from an economic point of view. This criterion is often used by development banks to determine credit worthiness of a proposed project.

International Centre for Water Hazard and Risk Management (ICWHRM) - An initiative of the Japanese government together with the Division of Water Sciences of UNESCO aimed at promoting sustainable development of river basins. The center, which should be formally approved by UNESCO's General Conference in 2005, will serve as an international hub for specific water sector problems based on the objectives of the IHP relating to risk management. The center will be located in Tsukuba, Japan.

International Conference on Financing for Development (ICFD) – A meeting of more than 50 heads of State and senior government officials held in Monterrey, Mexico in March of 2002 which may be considered as part of the run-up to the September 2002 World Summit on Sustainable Development. The meeting adopted the Monterrey Consensus and pledges were made by a number of developed countries to increase ODA.

International Conference on Population and Development – A conference sponsored by the United Nations to discuss global dimensions of population growth and change in Cairo, Egypt in September 1994. The conference marks the achievement of a new consensus on effective ways to slow population growth and improve quality of life by addressing root causes of unwanted fertility.

International Development Association (IDA) – The World Bank Group's concessional lending window which provides long-term loans at zero interest to the poorest of the developing countries. (WB)

International Energy Agency (IEA) – Formed in 1974, the IEA is an autonomous agency based in Paris that acts as the energy forum for the Organization for Economic Cooperation and Development. Members take joint action to meet oil supply emergencies, share energy information, coordinate their energy policies and cooperate in the development of rational energy programs. (IEA)

International Fund for Agricultural Development (IFAD) – The IFAD is a specialized agency of the United Nations established as an international financial institution in 1977 as a result of the 1974 World Food Conference that was organized in response to the food crises of the early 1970s. IFAD was created to mobilize resources on concessional terms for programmes that alleviate rural poverty and improve nutrition. (IFAD)

International law – The body of rules that govern the relationships of states with each other. International treaties are a part of international law. Sources of international law (from Article 38 of the Statute of the International Court of Justice) include: international conventions of general or particular nature; international custom, as evidence of a general practice accepted as law; and the general principles of law recognized by civilized nations.

International Monetary Fund (IMF) – The IMF is an international organization of 182 member countries, established to promote international monetary cooperation, exchange stability, and orderly exchange arrangements; to foster economic growth and high levels of employment; and to provide temporary financial assistance to countries to help ease balance of payments adjustment. (IMF)

International Pollinators Initiative (IPI) – The International Initiative for the Conservation and Sustainable Use of Pollinators is a crosscutting initiative within the program of work on agricultural biodiversity. (CBD)

International public good – Goods that will tend, in an international market economy, to be provided in quantities that are too low from the point of view of social efficiency. They are characterized by non-rivalry and non-excludability. If, for example only the Annex I-countries would carry out global greenhouse strategies, the non-Annex countries would benefit at no cost for themselves. The Annex I-countries cannot prevent the non-Annex countries from taking advantage of their greenhouse strategies, nor can they force them to pay for their investments. This may lead to a situation wherein a too low amount of the good - international commitments - will take place. (DFID)

International Tropical Timber Organization (ITTO) – An institution of producers formed in 1987 to promote conservation and sustainable development of tropical forests. (ITTO)

International Water Management Institute (IWMI) – One of sixteen centers supported by the Consultative Group on International Agricultural Research (CGIAR). Headquartered in Colombo, Sri Lanka, IWMI's mission is to improve water and land resources management for food, livelihoods and nature.

International waters – One of the seven focal areas of GEF defined as the seas, shared river and lake basins and shared estuaries and wetlands and shared groundwater aquifers. The distinguishing feature is that more than one nation has access to or makes use of them. (GEF)

International Whaling Commission (IWC) – The IWC was established under the International Convention for the Regulation of Whaling which was signed in Washington DC on 2 December 1946 Its purpose is to provide for the proper conservation of whale stocks and support the orderly development of the whaling industry. It is to review and revise as necessary the measures laid down in the Schedule to the Convention which governs the conduct of whaling throughout the world These measures provide for the complete protection of certain species; designate specified areas as whale sanctuaries; set limits on the numbers and size of whales which may be taken; prescribe open and closed seasons and areas for whaling; and prohibit the capture of suckling calves and female whales accompanied by calves.

Interpretation – A term referring to translating the written or spoken word into other languages. (MW)

Intra fauces terra – "In the jaws of the land." A principle for defining territorial seas. (BLD)

Intrinsic value – A form of judgment that takes into account all the values present in the system, an holistic valuation or fitness measurement of the whole. (CSG)

Introduced species – A species occurring in an area outside of its historically known natural range as a result of intentional or accidental dispersal by human activities.

Intuitive Knowledge - Knowledge that is not dependent on cognition (contact between "sense" and the "object") and can only be perceived after a person's mind is "silenced" through special meditative practices and attitudes. (Bharati).

Invasive alien species – A species of plant, animal or other organism that is (1) non-native (or alien) to the ecosystem under consideration and (2) whose introduction causes or is likely to cause economic or environmental harm or harm to human health. (CBD)

Ipso facto – By the fact (or act) itself. (BLD)

Island biogeography – The study of the relationship between island area and species number and between isolated continental areas which are effectively islands for many species.

ISO 9000 and 9001 – a set of standards of quality management (9000) and registration (9001) accepted by more then 40 countries including quality objectives, continued improvement, and monitoring of customer satisfaction.

ISO 14001 – A set of voluntary international standards that address consistency in environmental management systems, environmental auditing, environmental labeling, environmental performance evaluation, and the life cycle assessment.

Iteration – A loop that uses the current value of a system to derive its future value by re-inserting it into the equations controlling the system dynamics. (CSG)

IUCN – See **World Conservation Union**.

J

Jakarta Mandate – The Jakarta Mandate on Marine and Coastal Biological Diversity is a global consensus on the importance of marine and coastal biological diversity and is a part of the Ministerial Statement on the implementation of the CBD, as adopted by the COP/CBD held in Jakarta, Indonesia, in November of 1995. Through its programme of work, adopted in 1998, the Convention focuses on integrated marine and coastal area management, the sustainable use of living resources, protected areas, mariculture and alien species. (BCHM)

Johannesburg Declaration – The Johannesburg Declaration is one of the two documents agreed to at the 2002 World Summit on Sustainable Development. It reviews the current global challenges and the progress made in sustainable development from 1992 to 2002, commits is signatories to sustainable development, emphasizes the importance of working on a multilateral basis, and emphasizes the need to complete the work outlined in the WSSD "Plan of Implementation."

Johannesburg setting – A term used to identify ministerial level consultations and negotiation.

Joint contact group – Two contact groups created separately and brought together to resolve differences between them. (JG)

Joint Convention on the Safety of Spent Fuel Management and on the Safety of Radioactive Waste Management (2001) – This non-binding Convention seeks to achieve and maintain a high level of safety worldwide in spent fuel and radioactive waste management through the enhancement of national measures and international co-operation; to ensure that during all stages of spent fuel and radioactive waste management there are effective defenses against potential hazards so that individuals, society and the environment are protected from harmful effects of ionizing radiation now and in the future; and to prevent accidents with radiological consequences and to mitigate their consequences if they occur during any stage of spent fuel or radioactive waste management.

Joint implementation (JI) – A project-based activity in which one country can receive emission reduction credits when it funds a project in another country where the emissions are actually reduced. (UNFCCC)

Joint working group – Two working groups, each convened by a different body, brought together to work on cross cutting issues. (JG)

Jus civile – Law created within each country. (BLD)

Jus cogens – Compelling law. Peremptory principles of international law that cannot be overridden by specific treaties between countries; norms that are binding on all states at all times. (BLD)

Jus inter gentes – Law among peoples, nations. (BLD)

JUSSCANNZ – A coalition of the non-EU developed countries (and other invited countries) that acts as an information sharing and discussion forum. The acronym stands for Japan, US, Switzerland, Canada, Australia, Norway, and New Zealand.

K

Key – A term which typically means an item, idea or concept that is a deciding factor, i.e. is critical, instrumental or central to arriving at some determination.

Keystone species – A species that plays an important role in determining the overall structure and dynamic relationships within a biotic community. A keystone species presence is essential to the integrity and stability of a particular ecosystem.

Knock-on impacts – A term used primarily in the UK to mean secondary impacts.

Kyoto mechanisms – Formerly known as "flexibility mechanisms," the Kyoto mechanisms are based on market principles that Parties to the Kyoto Protocol can use in an attempt to lessen the potential economic impacts of GHG emission-reduction requirements. They include Joint Implementation (Article 6), the Clean Development Mechanisms (Article 12), and Emissions Trading (Article 17). (IPCC)

Kyoto Protocol (KP) – In 1997 at COP-3 in Kyoto, Japan the Parties adopted a protocol to the UNFCCC. The Kyoto Protocol calls for a reduction in GHG emissions by the industrialized nations and the nations with economies in transitions (Annex I countries) of at least 5% below 1990 levels (and in some cases 1995 levels) in the period 2008-2012. (IPCC)

L

L.docs – In-session documents that contain draft reports and texts for adoption by a COP or a subsidiary body. (UNFCCC)

Lacunae – Holes in the law; a gap or blank in a writing. (BLD)

Laissez faire – Free enterprise. The doctrine or system of government non-interference in the economy except as necessary to maintain economic freedom. (AD)

Land-based Sources Protocol (LBS) – The protocol concerned with land-based sources of pollution and polluting activities in the Wider Caribbean adopted 6 October 1999 in Oranjestad, Aruba as part of the Framework Plan of the Cartagena Convention. (UNEP)

Land mapping unit (LMU) – The smallest area of land that can be delineated on a map of a particular scale.

Languages, official UN – English, Spanish, French, Mandarin Chinese, Russian and Arabic.

La Niña – A period of stronger-than-normal trade winds and unusually low sea-surface temperatures in the central and eastern tropical Pacific Ocean; the opposite of *El Niño*. (NASA)

Large country – A country that is large enough for its international transactions to affect economic variables abroad, usually for its trade to matter for world prices. (AD)

LDC Fund – The LDC fund is intended to support the work program for the lesser-developed countries that "shall include, *inter alia*, National Adaptation Programmes of Action (NAPAs)." (UNFCCC)

LEAD – An international network of professionals committed to sustainable development. Established in 1991 by the Rockefeller Foundation, LEAD's "aim is to create and sustain a global network of emerging and working leaders ready and willing to address environment and development challenges in different countries." (LEAD)

Lead reviewer – An individual designated from a list of experts generally nominated by Parties because of their knowledge and experience in the topic to be covered in a report or other document to evaluate its accuracy and comprehensiveness.

Leapfrogging – A process by which, through technology transfer, developing countries adopt technologies designed and tested in industrialized countries without bearing the costs of research and development or experiencing the relatively slow stages of development that characterized the introduction of these technologies in the industrialized world. (UNEP)

Least developed countries (LDCs) – As defined by the United Nations, forty to fifty poor and vulnerable countries with an annual per capita income of less than US$ 1 per day: increasingly they work together as a coalition to defend their interests in international forums.

Least-cost alternative – An economic criterion applied in cost-benefit analysis in situations in which the benefits of projects or actions are equal. It then compares the net present value of the costs of the different projects. The alternative that has the lower present value of costs is preferred.

Legitimacy – The degree to which a government's procedures for making and enforcing laws are acceptable to the people. (UNDP)

Leipzig Declaration (1) – The full title is "The Leipzig Declaration on Conservation and Sustainable Utilization of Plant Genetic Resources for Food and Agriculture" which was made in Leipzig in 1996 at a conference organized by FAO. The meeting attended by delegates from 150 countries reached a consensus on urgent actions needed to protect the world's dwindling plant genetic resources for food and agriculture, funding, and the issue of Farmers' Rights. (FAO)

Leipzig Declaration (2) – A declaration made by a group of independent scientists researching atmospheric and climate problems concerned about the then upcoming climate conference in Kyoto, Japan in 1997 because it would impose "on citizens of the industrialized nations, but not on others – a system of global environmental regulations that include quotas and punitive taxes on energy fuels." (UNFCCC)

Less developed country (LDC) – A country where the annual per capita income is between $300 and $700. (UN)

Lessons learned – The notation and analysis of the achievements and errors of an activity that result in success or failures of material, timing or other mistakes, so that future similar activities may be improved.

Lex posterior derogat priori – More recent law prevails over (abrogates, overrules, trumps) an inconsistent earlier law. (BLD)

Liability and redress – A term describing the duty of reparation for damage when three conditions are met: (1) a breach of a legal obligation; (2) the breach can be attributed to an actor; and (3) causality can be established. There are at least two regimes: responsibility and liability. Under the first, the duty to repair exists only when a wrongful act is committed, and where wrongfulness is defined. Under the second, liability falls on the institution or person who benefits from or controls the activity even if there is no wrongfulness. Compensation may be paid by the liable actor through obligatory insurance, by a financial pool created by companies, and by the state of origin.

Life-cycle approaches – Strategies that help evaluate and address environmental issues and opportunities holistically with the aim of designing product service systems and reducing potential environmental impacts over the entire life cycle. (UNEP)

- Life-cycle assessment – An objective process for evaluating the environmental burdens associated with a product, process or activity by identifying the energy and materials used and wastes released to the environment, and to evaluate and implement opportunities to effect environmental improvements. (SETAC)
- Life-cycle management – An integrated framework of concepts, techniques and procedures to address environmental, economic, technological and social aspects of products and organizations, in order to achieve continuous environmental improvement from a life-cycle perspective. (SETAC)
- Design for environment – The systematic integration of environmental and social considerations into product and process design. (NRCC)
- Product service systems – A marketable mix of products and services that jointly fulfill client needs with less environmental impact. (UNEP)

- Integrated product policy – A process of working to stimulate each part of each phase of a product or service life-cycle to improve its environmental performance, from natural resources extraction through design, manufacture, assembly, marketing, distribution, sale, use, recycling (partial or complete) and disposal. (EU)

Little Ice Age – A cold period that lasted from about AD 1550 to about AD 1850 in Europe, North America, and Asia.

Living modified organisms (LMOs) – All organisms produced through the use of recombinant DNA technology, with a wider range of modifying technologies relevant when considering living modified prokaryotes and yeast. (BSWG)

Lomé Convention – An agreement originally signed in 1975 committing the European Union to programs of assistance and preferential treatment for the ACP Countries. The Cotonou Agreement replaced the Lomé Convention in June 2000. (AD)

London Club – An *ad hoc* forum for restructuring negotiations concerning debt owed to commercial banks. The forum is formed at the initiative of the debtor country and is dissolved when a restructuring agreement is signed. London Club "Advisory Committees" are chaired by a leading financial firm and include representatives from a cross-section of other exposed firms.

Low-income country – A country having an annual gross national product (GNP) per capita equivalent to $760 or less in 1998. The standard of living is lower in these countries; there are few goods and services; and many people cannot meet their basic needs. There are currently about 58 low-income countries with populations of 1 million or more. Their combined population is almost 3.5 billion. (WB)

M

Mainstreaming – A term originally coined in the 1970s to describe education for students with disabilities although now, in terms of GEG, it has come to mean "inclusive" (i.e. "mainstreaming environment" means to include environmental protection measures or environmental management considerations in the policies and programs of an institution more or less on the same level as economic and social considerations).

Majority – A number greater than half of a total.

Malmö Declaration – A ministerial level declaration (May 2000) which focuses on the major environmental challenges of the 21st century, the private sector, civil society, and a 10-year review of the UN Conference on Environment and Development.

Man and the Biosphere Programme (MAB) – An interdisciplinary programme guided by UNESCO dedicated to research and training intended to develop the basis for the rational use and conservation of the resources of the biosphere.

Manila Declaration – The full name for this declaration is "The Manila Declaration Concerning the Ethical Utilization of Biological Resources" which was developed at the ASOMPS meeting in Manila, Philippines in February of 1992. (BCHM)

Mare clausum – Closed seas as opposed to *mare liberum* (freedom of the seas). (BLD)

Mare liberum – Freedom of the seas as opposed to *mare clausum* (closed seas). (BLD)

Mediation – The bringing about a peaceful settlement or compromise between parties to a dispute through the benevolent intervention of an impartial third party.

Medium-sized projects – Medium-sized projects require no more than US$1 million in GEF financing for the preparation and implementation. Most GEF projects average $5.5 million per project and take several years to implement. As the GEF has gained experience in implementing projects, it has been recognized that smaller projects could benefit from expedited procedures so that they could be designed and executed more quickly and efficiently. (GEF)

Meeting of the Parties (MOP) – Whereas the Conference of the Parties carries with it the sense of "association" or "members" and is reserved for conventions, "meeting of the parties" is meant to convey the sense of an encounter and is reserved for a meeting of the parties to a protocol.

Mekong River Commission (MRC) – A commission established in 1995 by the governments of Cambodia, Laos, Thailand and Vietnam to help manage the 4,880 km river. Myanmar and China are not members although the river flows through their territories.

Member states – Governments that officially participate in an intergovernmental organization or a "party" to a convention.

Memorandum of understanding (MOU) – An agreement defining the roles and responsibilities of all parties in relation to the other parties on an issue of interest to each (often made between International organizations). (UNT)

Mercosur – The "common market of the South," created by the 1991 Treaty of Asunción. Mercosur originally included Argentina, Brazil, Paraguay, and Uruguay. Chile and Bolivia became associate members in 1996 and 1997, respectively. (AD)

Methane – A hydrocarbon with four hydrogen atoms attached to each carbon atom (CH_2).

Middle-income country – A country having an annual gross national product (GNP) per capita equivalent to more than $760 but less than $9,360 in 1998. The standard of living is higher than in low-income countries, and people have access to more goods and services, but many people still cannot meet their basic needs. There are currently about 65 middle-income countries with populations of one million or more. Their combined population is more than 1.5 billion. (WB)

Millennium Declaration – The United Nations Millennium Declaration was adopted at the Millennium summit in September 2000 by 147 heads of State and Government, and 191 nations in total. The Declaration spells out values and principles, as well as goals in the priority areas of peace, development (reducing global poverty by half and creating universal primary education by the year 2015), the environment, human rights, protecting the vulnerable, the special needs of Africa, and strengthening the UN. (UNMD)

Millennium Development Goals (MDGs) – The Millennium Declaration, adopted by the UN member states in 2000, establish eight Millennium Development Goals to be reached by the year 2015:

- Eradicate extreme poverty and hunger;
- Achieve universal primary education;
- Promote gender equality and empower women;
- Reduce child mortality;
- Improve maternal health;
- Combat HIV/AIDS, malaria and other diseases;
- Ensure environmental sustainability; and,
- Develop a global partnership for development

The goals detail eighteen specific development targets, each of which has, a time frame, and indicators designed to monitor degree of achievement. The cost to attain these goals is between $40-$60 billion each year in additional assistance. (UN)

Millennium Ecosystem Assessment (MA) – A programme modeled on the Intergovernmental Panel on Climate Change as a multi-scale, interdisciplinary assessment of biodiversity to meet the scientific needs of the Ramsar Convention on Wetlands, the Convention on Biological Diversity (CBD) and the Convention to Combat Desertification.

Minister (1) – A high officer of a state entrusted with management of a division of government activities. (MW)

Minister (2) – A diplomatic representative ranking below an ambassador. (MW)

Misc. docs – Miscellaneous documents issued on plain paper with no masthead: they generally contain views or comments submitted as received from a delegation without formal editing.

Mission - Official travel for an administration, representational or technical purpose. (UN)

Mitigation – Actions taken during the planning, design, construction and operation of works and undertakings to alleviate the potential adverse effects of those activities on other interests.

Modality – A ceremonial form, protocol, or condition that surrounds formal agreements or negotiations. (Dc)

Modus vivendi – An instrument recording an international agreement of temporary or provisional nature intended to be replaced by an arrangement of a more permanent and detailed character. It is usually made in an informal way, and never requires ratification. (UNT)

Monitoring – A continuous process of collecting and analyzing information to measure the progress of a project toward expected results. Monitoring provides managers and participants with regular feedback that can help determine whether a project is progressing as planned. (UNDP)

Monitoring and evaluation (M&E) – The periodic evaluation of management activities to determine how well objectives were met and how management practices should be adjusted.

Monterrey Consensus – The primary document resulting from the ICFD that has as its goal to "eradicate poverty, achieve sustained economic growth and promote sustainable development" that looks toward a "fully inclusive and equitable global economic system." Actions are to treat the issues of the mobilization of domestic financial resources for development; mobilization of international resources for development, foreign direct investment and other private flows; international trade as an engine for development; the augmentation of international financial and technical cooperation for development; external debt; and economic issues, and the coherence and consistency of international monetary, financial and trading systems in support of development.

Montevideo Programme – A long-term strategic programme of UNEP that provides the foundation for the development of environmental law. Its achievements include the development of global conventions on the ozone layer, transboundary movements of hazardous wastes, biological diversity, and information exchange on hazardous chemicals.

Montreal Protocol (on Substances that Deplete the Ozone Layer) – The principal international agreement under which ozone-depleting compounds are regulated. Established in 1987 and entered into force on 1 January (1992), it has subsequently been adjusted and amended. (UNT)

MOSAICC – A voluntary code of conduct developed to facilitate easy access to, and international circulation of, microbial genetic resources for the benefit of science and worldwide sustainable development in the framework of the Convention on Biological Diversity. (BCHM)

Most favored nation clause – A principle that guides progress towards trade liberalization within GATT and WTO whereby any advantage granted by one signatory country to another must be extended to all the other signatory countries. (AD)

Multilateral Agreement on Investments (MAI) – The Multilateral Agreement on Investments is the result of negotiations within the Organization for Economic Cooperation and Development and is designed to eliminate nearly all barriers to international trade by liberalizing international trade laws and protecting foreign investments. Its basic provisions are to open most economic and natural resource sectors to foreign ownership; treat foreign firms fairly and equally; remove restrictions against the movement of capital; allow individual firms to sue foreign governments before an international mediation panel; and give full and proper compensation for expropriation. It was never adopted. (TWN)

Multilateral aid – Aid channeled through multilateral institutions for use in, or on behalf of, aid recipient countries.

Multilateral institutions – International institutions with governmental membership, spanning several regions, including financial institutions such as the World Bank and IMF, UN agencies and regional groupings.

Multi-stakeholder (MS) – Term used to describe the case where there are a number of parties interested in the outcome of a decision; *Agenda 21* lists women, youth, Indigenous Peoples, NGOs, business and industry, workers and trade unions, science and technology, farmers, local authorities as being the major stakeholders that should be included in environment/development discussions. (UNED)

Multi-stakeholder dialogue – Meetings that enable direct interactions between governments and major groups on specific topics and as part of the official inter-governmental process rather than taking place on the margins. The dialogues allow major groups to discuss their concerns, experiences and proposals in specific areas in detail with governments.(WSSD)

Mutates mutandis – The necessary changes having been made; having substituted new terms; with respective differences taken into consideration.

M

Mutually coherent policies – Policies that are related to one another such that they are consistent, cohesive and understandable.

N

National delegation – One or more officials who are empowered to represent and negotiate on behalf of their government. (BCHM)

National Environment Action Plan (NEAP) – A Plan that identifies key environmental problems, sets priorities for dealing with them and leads to a comprehensive national environmental policy and programs to implement the policy. (GEF)

National income accounts – A system of accounting used to measure a nation's economy, the results of which are often given in terms of GNP or GDP.

National interest, concept of – A concept that identifies interests of one country that are distinct from or even contrary to those of another.

National Strategy for Sustainable Development (NSSD) – According to an agreement produced at the UN Rio+5 Conference, the NSSD is a plan of national action developed to achieve sustainable development goals.

Native species – Plants, animals, fungi, and microorganisms that occur naturally in a given area or region.

Natural law – A theory based on the premise that there are certain normative principles that are "self evident" and which exist independently of their codification or enforcement by human beings. (ECO)

Natural resource – A term describing anything that is provided by nature, such as minerals deposits, forests, water, wildlife, etc. The term is being replaced by "natural" or "ecosystem services."

Negative sum – The idea, from game theory that agents combine in such a way that both lose or that the total change is a reduction in overall fitness, sometimes called dysergy or "lose-lose".

Negotiated science – Conclusions of government appointed scientists that are criticized by non-governmental scientists as representing compromises rather than pure scientific findings.

Negotiating state – A state that took part in the drawing up and adoption of the text of the treaty. (VC)

Negotiation – A process whereby two or more people who begin with conflicting positions attempt to reach an agreement by modifying their original positions or by developing new proposals that reconcile the interests underlying them. The preliminary discussions leading up to the adoption of an agreement.

Neoclassical economics – A system of economics that seeks the optimum allocation of resources through market competition in which the well being of one individual cannot be increased without that of another being diminished.

Neoliberalism – A business-government philosophy based on the primacy of the individual with only a minimal role for government. Neoliberal policies recommend solutions based on free movement of goods, services and capital, with reliance on market forces to allocate resources.

Net present value (NPV) – An economic criterion used to evaluate whether a project is economically viable or not where the discounted value of the expected costs of an investment is deducted from the discounted value of the expected benefits of the same investment. If the NPV is positive the project or action is usually considered worthwhile. Likewise the NPVs of more than one project can be compared to isolate the one that would theoretically be preferred.

No net loss – A working principle that strives to balance unavoidable habitat losses with habitat replacement on a project-by-project basis.

No-regrets measures – Decisions and actions taken which can reduce greenhouse gas emissions but which merit implementation purely on a financial basis.

Non-consumptive value – A value assigned to a resource that is not diminished by the use of that resource.

Non-governmental actor – The full range of NGOs and civil society including the major groups identified at UNCED.

Non-governmental organization (NGO) – A non-profit group or association organized outside of institutionalized political structures to realize particular social objectives or serve particular constituencies. (GBS)

Non-group – Under circumstances of extreme reluctance to enter into negotiation, non-groups can be called into being by the Chair in order to encourage communication without the pressure of negotiations. (JG)

Non-linear – A term used to describe systems that behave in unexpected ways and that do not change proportionally to a change in input. The change may sometimes go down when you expect it to go up, do nothing instead, or change drastically with only minor changes to the input. (CSG)

Non-methane volatile organic compounds (NMVOC) – Compounds generally emitted from combustion sources that can combine with other pollutants to produce photochemical smog.

Non-papers – In-session documents issued informally to facilitate negotiations; they do not have an official document symbol or number although they may have an identifying number or the name of the author(s). (CBD)

Non-party – A state that has not ratified a convention but may attend meetings of a COP, MOP or their subsidiary bodies as an observer.

North American Biodiversity Information Network (NABIN) – An initiative of Mexico, the United States and Canada to provide greater access to biodiversity information.

North American Free Trade Agreement (NAFTA) – An agreement implemented in 1994 committing Canada, the U.S., and Mexico to the elimination of all tariffs, quotas, and other trade barriers between them before 2009.

Northern countries – High-income or developed countries.

North-South – A term indicating collaboration or technology sharing between a "northern" country and one from the "south."

Notification – A formality through which a state or an international organization communicates certain facts or events of legal importance. Instead of opting for the exchange of documents or deposit, states may be content to notify their consent to the other party or to the depositary. (VC)

O

Objection – Any signatory or contracting state has the option of objecting to a reservation, *inter alia*, if, in its opinion, the reservation is incompatible with the purpose of the treaty. (VC)

Observer – A state or other institution such as the specialized agencies of the UN, other international organizations, and qualified governmental and non-governmental organizations that are accredited to attend the meetings of COP, MOP and their subsidiary bodies.

OECD – See **Organization for Economic Cooperation and Development**.

Official Development Assistance ODA – The amount that a nation expends through grants and other development assistance programs calculated as a percent of GNP.

Offsetting effect – Increased consumption resulting from actions that increase efficiency and reduce consumer costs. One example would be a home insulation program that reduces heat loss by 50% but does not bring about an equal reduction in energy consumption, because residents can then afford to keep their homes warmer more cheaply. (VTPI)

Oil Spills Protocol – The protocol concerning cooperation in combating oil spills in the Wider Caribbean Region that was adopted and entered into force concurrently with the Cartagena Convention. (UNEP-CAR/RCU)

Open-ended Intergovernmental Group of Ministers or their Representatives (IGM) – The group of ministers mandated in Decision 21/21 of the UNEP Governing Council (February 2001) to undertake a comprehensive policy oriented assessment of existing institutional weaknesses as well as future needs and options for strengthened international environmental governance, including the financing of the United Nations Environment Programme. (UNEP)

Operational focal point – A focal point charged with coordinating GEF operational matters within a country. (See **Political Focal Point**).

Opportunity cost – The value of goods or services foregone, including ecosystem goods and services.

Organization – A social group with a structure designed to achieve collective goals. (UNDP)

Organization for Economic Cooperation and Development (OECD) – An intergovernmental organization that includes most of the world's industrialized, market economies (United States, Austria, Belgium, Canada, Denmark, Finland, France, Germany, Greece, Iceland, Ireland, Italy, Luxembourg, Netherlands, Norway, Portugal, Spain, Sweden, Switzerland, Turkey, United Kingdom, Japan, Australia, and New Zealand).

Organization of African Unity (OAU) – The Organization of African Unity was established in 1963 to promote self-government, respect for territorial boundaries, and social progress throughout the African Continent. Membership is open to all independent African countries. The OAU was replaced in 2001 by the African Union.

Organization of American States (OAS) – A western hemisphere organization established in 1890 (as the Pan American Union) to provide a forum for political, economic, social, and cultural cooperation among its 35 member and 54 observer states. The OAS Charter was opened for signature in 1948 and entered into force in 1950.

Overarching role – A term variously used to mean the "preponderant," "strategic," and/or "overriding" objective or theme of an institution that may include a sense of a high degree of formal and informal influence.

Oxfam – A British NGO formed in 1942 as a development, relief, and campaigning organisation dedicated to finding lasting solutions to poverty and suffering around the world. (Ox)

P

Pacta sunt servanda – The doctrine that agreements must be observed (that is: honored, and obeyed). (BLD)

Pan American Health Organization (PAHO) – A specialized organization of the Organization of American States and a regional agency of the World Health Organization founded in 1902. It provides technical assistance to Latin America and the Caribbean in environmental health and sanitation; water supply, wastewater, control of air, water and soil pollution and hazardous wastes. (PAHO)

Pan American Union – The predecessor to the OAS and the world's oldest regional organization dating to the First International Conference of American States held in Washington, DC from October 1889 to April 1890.

Paris Club – A group of 19 governments that hold claims on other governments either directly or through their government institutions. Created to maintain strict standards for negotiating debt-restructuring arrangements with bilateral creditors. The following countries are permanent members: Austria, Australia, Belgium, Canada, Denmark, Finland, France, Germany, Ireland, Italy, Japan, The Netherlands, Norway, Russian Federation, Spain, Sweden, Switzerland, United Kingdom, and the United States of America.

Participation – A process through which stakeholders influence and share control over development initiatives and the decisions and resources which affect them. It is a process which can improve the quality, effectiveness and sustainability of projects and strengthen ownership and commitment of government and stakeholders. (WB)

Participatory urban governance – The relative degree to which citizens are involved in local policy development, planning, programming, budgeting, service delivery and monitoring. (UNCHS)

Partnership (1) – A relationship resembling a legal partnership and usually involving close cooperation between parties having specified and joint rights and responsibilities. (MW)

Partnership (2) – Two or more parties combining knowledge, skills and means to achieve a common goal. (UNESCO)

Party – A State that has consented to be bound by the treaty and for which the treaty is in force. (VC)

Patent – An incorporeal statutory right that gives an inventor, for a limited period, the exclusive right to use or sell a patented product, method or process. (MW)

PDF A, PDF B – See **Project Preparation and Development Facility**. (GEF)

People's Summit – A series of meetings held in Rio de Janeiro in 1992 to coincide with the meetings of UNCED and that were attended by thousands of private citizens, and representatives from NGOs, indigenous peoples, businesses, scientific and spiritual groups from all over the world.

Perdiem - See Daily subsistence allowance (UN)

Persistent organic pollutants (POPs) – Chemical substances that bioaccumulate through the food web and pose a long-term risk to human and environmental health.

PIC Convention – The PIC Convention (The Rotterdam Convention on the Prior Informed Consent Procedure for Certain Hazardous Chemicals and Pesticides in International Trade) was opened for signature in 1998 and will come into force when ratified by 50 countries. The Convention will allow monitoring and control of international trade in various health threatening chemicals. Importing countries will decide which chemicals they want to receive and to exclude those they cannot safely manage. (PIC)

Plan of action – An official document which outlines the goals, objectives, mandates, partnerships, coordination and management processes, time frame and funding mechanisms of an initiative, project or program.

Plan of Implementation – The Plan of Implementation is one of two documents signed at the 2002 World Summit on Sustainable Development. It is a framework for action for implementation of the UNCED commitments and includes an introduction and chapters on poverty eradication; consumption and production; the natural resource base; health; small island developing States (SIDS); Africa; other regional initiatives; means of implementation; and an institutional framework.

Plant genetic resources for food and agriculture (PGRFA) – Any material of plant origin, including reproductive and vegetative propagating material, as well as the genetic parts and components, containing functional units of heredity, of actual or potential value for food and agriculture. (FAO)

Plenary – An open session of the entire Conference of the Parties to a convention or to any of the subsidiary bodies where all formal decisions are made. (BCHM)

Plenipotentiary – A person who has full power to do something. The person authorized by an instrument of full powers to undertake a specific treaty action. (UNT) In international law, a senior diplomatic officer commissioned to act for his or her government. A *minister plenipotentiary* is one who by custom ranks immediately below an ambassador and is of the same rank as an envoy extraordinary.

P-level staff – A term used to categorize professional level staff (and salary/benefit package) in the UN and OAS systems, starting at P1 and rising to P5.

Political focal point – A focal point responsible for coordination matters related to GEF governance. (See **Operational focal point**).

Political will – See "C.H.A.O.S.S. on Chaos" in the essay.

Polity – A specific form of political organization. (MW) The form of government of a nation, state, or organization or the political organization itself, such as a nation. (GC)

Polluter Pays Principle – The concept that the cost of measures needed to improve a polluted environment to an acceptable state should be reflected in the cost of the goods and services whose production (or consumption) caused the pollution.

Pollution prevention – The use of processes, practices, materials, products or energy that avoid or minimize the creation of pollutants and waste, and that reduce the overall risk to human health or the environment. Pollution prevention and cleaner production both focus on a strategy of continuously reducing pollution. (EC)

Population – A group of individuals with common ancestry that are much more likely to mate with one another than with individuals from another such group. (GBS)

Portfolio – The area of official responsibility or duties of an individual, office or institution; the collection of programs and projects belonging to a given theme.

Positive sum – The idea, from game theory that when agents interact they can both benefit, the whole being greater than the sum of the parts; also called synergy or "win-win." (CSG)

Precautionary Principle – A principle adopted by the 1992 UN Conference on Environment and Development (Rio principle 15) that says (1) that environmental protection measures should be applied where threats of serious or irreversible environmental damage can be seen even though there is a lack of full scientific certainty of the threat and (2) that this lack certainty should not be used to postpone cost-effective actions to prevent environmental degradation. Used for the first time at the 1990 Bergen Conference which stated: "when confronted with serious or irreversible environmental threats, the absence of absolute scientific certainty should not serve as a pretext for delaying the adoption of measures to prevent environmental degradation." Also referred to as the "precautionary approach."(UNFCCC)

Preparatory committee (PrepCom) – A committee established as an administrative step in the process to develop an international conference in order to achieve consensus on issues and pre-negotiated documents that world leaders are to approve.

President (1) – Elected individual who presides over a meeting or assembly. For example, a person, generally a senior official or minister from the state or region hosting the meeting elected by the COP to preside over that meeting. (MW)

President (2) – An elected official having the position of both head or chief of state and chief political executive in a republic having a presidential government. (MW)

President (3) – An elected official having the position of head or chief of state, but with only minimal or ceremonial political powers in a republic having a parliamentary government. (MW)

Primary productivity – The rate at which new plant biomass is formed by photosynthesis. Gross primary productivity is the total rate of photosynthetic production of biomass; net primary productivity is gross primary productivity minus the respiration rate. (NASA) The transformation of chemical or solar energy to biomass. Most primary production occurs through photosynthesis. (GBS)

Prime Minister – The chief political executive in a republic having a parliamentary government.

Prior informed consent, principle of (PIC) – A principle established to help participating countries know the characteristics of potentially hazardous chemicals that may be shipped to them. The aim is to promote a shared responsibility between exporting and importing countries to protect human health from the harmful effects of certain hazardous chemicals that are traded internationally. (PIC)

Private sector – In a mixed economy, the part of the economy not under government control and that functions within the market; private enterprise. (UNDP)

Privatization – Policies that allow the change of a business, industry, activity, or institution from public to private control or ownership. (AD)

Procedural equity, principle of – A principle that refers to how decisions are made: (a) equity implies that those who are affected by decisions should have some say in the making of these decisions and (b) equity must ensure equal treatment before the law.

Process indicators – Monitoring tools that can provide information about where interventions are needed. (UNFPA)

Program – A coherent and interactive set of proposals and projects that are synchronized within a planning and implementation scheme.

Programme – A portfolio of projects that aims to achieve a strategic goal of an organization, planned and managed in a coordinated way.

Programme of Action of the International Conference on Population and Development (IPCD) – The (ICPD) was held from 5-13 September 1994 in Cairo, Egypt. ICPD delegates considered the interconnectedness of population, poverty, gender, patterns of production and consumption, and the environment. The 1994 Programme of Action recommended a set of important population and development objectives, including: sustained economic growth in the context of sustainable development; education, especially for girls; gender equity and equality; infant, child, and maternal mortality reduction; and the provision of universal access to reproductive health services, including family planning and sexual health care.

Programme of Action of the Special Session of the General Assembly to Review and Appraise the Implementation of *Agenda 21* (UNGASS) – The 19th special session of the UNGA convened in June of 1997 (Earth Summit +5) to review the outstanding issues not fully resolved at the UNCED including finance, climate change, freshwater management, and forest management. This session adopted the Programme for the Further Implementation of *Agenda 21* (A/RES/S-19/2). It assessed progress since UNCED, examined implementation, and established the CSD's work programme for the period 1998-2002, leading up to WSSD.

Project – A planned and goal-oriented socio-economic development activity requiring financial investment or human participation over a given time.

Project cycle – An ordered series of activities that include planning, approval, execution, monitoring, and evaluation.

Project Preparation and Development Facility (PDF) – Facility that grants money to the implementing agencies for developing project concepts in three blocks: Block A (up to $25,000), Block B (up to $750,000) and mid-size or Block C (up to $1 million). Block A funds are for pre-project activities at national level. Block B funds are for obtaining information to complete project proposals. Block C funds are a one-time

GEF grant for completing technical design and feasibility studies for large-scale projects to be funded by another entity. (GEF)

Protected area – A geographically defined area (land or sea) that is designated or regulated and managed to achieve specific conservation objectives (CBD). Some definitions suggest that the area needs to be legally designated. (WRI)

Protected Area Management Categories (IUCN) – A set of management objectives assigned to ten different classifications of protected areas:

- Category I: *Scientific Reserve/Strict Nature Reserve*. To protect nature and maintain natural processes in an undisturbed state in order to have ecologically representative examples of the natural environment available for scientific study, environmental monitoring, education, and for the maintenance of genetic resources in a dynamic and evolutionary state.
- Category II: *National Park*. To protect natural and scenic areas of national or international significance for scientific, educational and recreational use.
- Category III: *Natural Monument/Natural Landmark.* To protect and preserve nationally significant natural features because of their special interest or unique characteristics.
- Category IV: *Managed Nature Reserve/Wildlife Sanctuary*. To assure the natural conditions necessary to protect nationally significant species, groups of species, biotic communities, or physical features of the environment where these require specific human manipulation for their perpetuation.
- Category V: *Protected Landscape or Seascape*. To maintain nationally significant natural landscapes which are characteristic of the harmonious interaction of man and land while providing opportunities for public enjoyment through recreation and tourism within normal life style and economic activity of these areas.
- Category VI: *Resource Reserve*. To protect resources of the area for future use and prevent or contain development activities that could affect the resource pending the establishment of objectives which are based upon appropriate knowledge and planning.
- Category VII: *Natural Biotic Area/Anthropological Reserve*. To allow the way of life of societies living in harmony with the environment to continue undisturbed by modern technology.

- Category VIII: *Multiple-Use Management Area/Managed Resource Area*. To provide for the sustained production of water, timber, wildlife, pasture, and outdoor recreation, with the conservation of nature primarily oriented to the support of economic activities (although specific zones may also be designed within these areas to achieve specific conservation objectives).
- Category IX: *Biosphere Reserve*. These are part of an international scientific programme, the UNESCO Man and the Biosphere (MAB) Programme, which is aimed at developing a reserve network representative of the world's ecosystems to fulfill a range of objectives, including research, monitoring, training and demonstration, as well as conservation roles.
- Category X: *World Heritage Site*. The Convention Concerning the Protection of the World Cultural and Natural Heritage (which was adopted in Paris in 1972 and came into force in December 1975) provides for the designation of areas of "outstanding universal value" as World Heritage Sites, with the principal aim of fostering international cooperation in safeguarding these important sites.

Protocol – An international agreement supplementary to a convention that adds to or changes some provision of the convention only for the states parties who adopt the protocol. (VC)

Prototype Carbon Fund (PCF) – World Bank funding of $145 million US to Uganda, Chile, and Latvia for these and other developing countries to adapt to climate impacts and develop emissions reducing technologies. Under this fund, the World Bank will buy carbon dioxide emissions reductions for up to $3.9 million over 15 to 20 years.

Provinciality effect – Increased diversity of species because of geographical isolation. (IUCN)

Public – One or more natural or legal persons, and, in accordance with national legislation or practice, their associations, organizations or groups. (UNECE)

Public authority – A government at the national, regional or other level; natural or legal persons performing public administration functions under national law, including specific duties, activities or services; other natural or legal persons having public responsibilities or functions, or providing public services; and, the institutions of any regional economic integration organization. (UNECE)

Public concerned – The public affected or likely to be affected by, or having an interest in, environmental decision-making. Non-governmental organizations promoting environmental protection and meeting any requirements under national law shall be deemed to have an interest. (UNECE)

Public consultation – The process of engaging affected people and other interested parties in open dialogue through which a range of views and concerns can be expressed in order to inform decision-making and help build consensus. (WB)

Public good – A good that is impossible to keep anyone from using and where one person's consumption of that good does not reduce the quantity available for use by someone else.

Public participation – A process by which the views and concerns of the public are identified and incorporated into decision-making by public institutions. Participation is a process through which stakeholders influence and share control over development initiatives and the decisions and resources which affect them. Includes access to information, access to process and access to justice.

Public sector – The part of the economy that is not privately owned, either because it is owned by the state or because it is subject to common ownership. Includes the national government, local authorities, national industries and public corporations. (UNDP)

Punta del Este Declaration – A 1986 declaration from the parties to the General Agreement on Tariffs and Trade (GATT) written to launch the Uruguay Round of multilateral trade negotiations. (AD)

Q

Qualified majority – The raising of the simple majority requirement of "50 percent plus one" to a higher level, in order to protect the rights of the minority; thus, a system of weighted votes. For example for certain decisions in the EU, Member States' votes are weighted based on their population and corrected in favor of less-populated countries.

Quality assurance/quality control (QA/QC) – The process of strengthening research outcomes by applying rigorous and systematic approaches to planning and conducting scientific investigations.

Quod hoc – On this matter.

Quorum – The minimum number of people/governments/institutions required to formally open a meeting.

R

Ramsar Convention – Convention on Wetlands of International Importance especially as Waterfowl Habitat, signed at Ramsar, India on 2 February 1971. It came into force on 21 December 1975. (RC)

Rapporteur – The officer specifically charged with keeping the minutes and writing the report of each session of a meeting.

Rare – An IUCN category (*R*) of taxa with small world populations that are not at present Endangered or Vulnerable, but are at risk. These taxa are usually localized within restricted geographical areas or habitats or are thinly scattered over a more extensive range.

Ratification – An international act whereby a state, through congressional or parliamentary approval, indicates its consent to be bound to a treaty if the parties intended to show their consent by such an act (VC).

Rebound effect – See "Offsetting effect."

Recommendation – A determination that is weaker than a decision or a resolution and not binding on Parties.

Referendum – A vote by all voters on a question.

Reforestation – The action of renewing or improving forest cover by planting seeds or seedlings on land that had been previously forested.

Refugee – A person who, owing to well-founded fear, or fact, of being persecuted for reasons of race, ethnicity, religion, nationality, membership of a particular social group or political opinion, is outside the country of his/her nationality and is unable or, owing to such fear, is unwilling to avail him/her self of the protection of that country; or who, not having a nationality and being outside the country of former habitual residence as a result of such events, is unable or, owing to such fear, is unwilling to return to it. (UNDP)

Regime – An institutional arrangement formally or informally established to study and support a common property issue area (atmosphere, oceans, biodiversity, Antarctica, outer space, etc.).

Region – An area of territory that can be defined for spatial analysis and planning purposes as distinct and coherent. Planning regions can be administrative following the borders of particular jurisdictions within countries; or functional which follow the boundary o f the issue subject to the planning process, for example, river catchment areas and travel to work areas.

Regional Economic Integration Organization – An organization constituted by sovereign States of a given region, to which its member States have transferred competence in respect of matters governed by this Convention and which has been duly authorized, in accordance with its internal procedures, to sign, ratify, accept, approve or accede to it. (CBD)

Regional group – One of five regional groupings within the UN system that meet to discuss issues and nominate Bureau members and other officials (Africa, Asia, Central and Eastern Europe (CEE), Latin America and the Caribbean (GRULAC), and WEOG, the Western Europe and Others Group.

Regional Seas Conventions and Action Plans – A series of 19 regional conventions and action plans administered by the UNEP (six programs) and other independent partner organizations (13 programs). Those of UNEP include:

- Barcelona Convention for the Protection of the Marine Environment and the Coastal Region of the Mediterranean (1976),
- Abidjan Convention for Cooperation in the Protection and Development of the Marine and Coastal Environment of the West and Central African Region (1981),
- East Asian Seas Convention and Action Plan (1981),
- Cartagena Convention for the Protection and Development of the Marine Environment of the Wider Caribbean Region (1983),
- Nairobi Convention for the Protection, Management and Development of the Marine and Coast Environment of the Eastern Africa Region (1985).

Those administered by other intuitions include:

- Helsinki Convention on the Protection of the Marine Environment of the Baltic Sea Area (1992),
- OSPAR Convention for the Protection of the Marine Environment of the North-East Atlantic (Oslo and Paris conventions, adopted 1974, revised and combined 1992),
- Arctic Council for the Protection of the Arctic Marine Environment (1992),
- Antarctic Treaty System: Convention for the Conservation of Antarctic Marine Living Resources (1982) and Madrid Protocol on the Protection of the Antarctic Environment (1991),
- Kuwait Regional Convention for Co-operation on the Protection of the Marine Environment from Pollution (1978),
- Lima Convention for the Protection of the Marine Environment and Coastal Area of the South-East Pacific (1981),
- Jeddah Regional Convention for the Conservation of the Red Sea and Gulf of Aden Environment (1982),
- Noumea Convention for the Protection of Natural Resources and Environment of the South Pacific Region (1986),
- Convention and Action Plan for the Caspian Sea and Bucharest Convention on the Protection of the Black Sea Against Pollution (1992),
- North-West Pacific Action Plan (1994),
- North-East Pacific Action Plan, and the,
- South Asian Seas Action Plan (1995).

Regional Seas Programme – The Regional Seas Programme was initiated in 1974 as a global programme implemented through regional components. The Programme at present includes thirteen regions involving more than 140 coastal States and Territories. (UNEP)

Regular funds/budget – Assessed contributions from member states of an international organization. (UNDP)

Regulations – Rules or orders issued by an executive authority or regulatory agency of a government and having the force of law.

Renewables – Energy sources that are, within a short timeframe relative to the earth's natural cycles, sustainable, and include non-carbon technologies such as solar energy, hydropower, geothermal, wave and wind as well as carbon-neutral technologies such as biomass. (IPCC)

Report of the World Food Summit (2002) – This report summarized the progress made in reaching the goals of the Rome Declaration on World Food Security and the World Food Summit POA. In all aspects it is one of the most criticized meetings ever held under the auspices of the United Nations.

Report of the World Summit for Social Development (WSSD) – The WSSD was held in March 1995 in Copenhagen, Governments reached a new consensus on the need to put people at the centre of development. In the final report governments pledged to make the conquest of poverty, the goal of full employment and the fostering of social integration overriding objectives of development.

Request for proposal (RFP) – A document issued by one organization to another organization (or to several other organizations) describing work that the issuer wishes to have undertaken by the recipient(s) and inviting the recipient(s) to respond with a proposal.

Reservation – A unilateral statement, however phrased or named, made by a State, when signing, ratifying, accepting, approving or acceding to a treaty, whereby it purports to exclude or to modify the legal effect of certain provisions of the treaty in their application to that State. (VC)

Reservoir (1) – A component or components of the climate system where a greenhouse gas or a precursor of a greenhouse gas is stored. (UNFCCC)

Reservoir (2) – A body of water created artificially by the construction of a dam across a flowing river or stream.

Resolution – A formal action such as a recommendation that does not normally become part of the body of decisions that guide the work of a COP. They are directives that guide rather than permanent legal acts. Formal resolutions of a general assembly are binding decisions.

Resolution 47/191 – A resolution of the UN General Assembly adopted in 1992 relative to the creation of the UN Commission on Sustainable

Development (CSD). The resolution presents the terms of reference, makeup, guidelines for NGO participation, organization of work, relations with other UN bodies and Secretariat agreements for the CSD.

Resolution 55/199 – A resolution of the UN General Assembly adopted in December 2000 that calls for a ten-year review of UNCED (September of 2002) at the summit level to reinvigorate global commitment to sustainable development. The review was to focus on accomplishments, identify areas requiring further efforts to implement *Agenda 21* and other UNCED outcomes, lead to action-oriented decisions, and result in renewed political commitment to achieve sustainable development.

Restoration – In ecosystem management restoration is the return of that system to some previous desired state or earlier successional state in terms of its principle components of species and functions.

Results – A broad term used to refer to the effects of a project. The terms "outputs," "outcomes" and "impact" describe more precisely the different types of results: (GEF)

- *Outputs*: tangible products (including services) of a project that are necessary to achieve its objectives.
- *Outcomes*: results of a project relative to its immediate objectives that are generated by the project outputs.
- *Impact*: results of a project that are assessed with reference to the development objectives or long-term goals of that project; changes in a situation, whether planned or unplanned, positive or negative, that a project helps to bring about.

Right to development – According to the 1986 UNGA 'Declaration on the Right to Development." The right to development " . . . is an inalienable human right by virtue of which every human person and all peoples are entitled to participate in, contribute to, and enjoy social, cultural and political development, in which all human rights and fundamental freedoms can be fully realized." Development, as a "right" is not accepted by all states.

Right to environment – A phrase that generally relates to the Rio Declaration and its principles which defined the rights and responsibilities of people to a clean and safe environment. The eighteen principles included in the Declaration provide substantial detail as to what the right includes and what responsibilities it implies. (UNCED)

Right to participation – A right generally referring to Articles 19, 20 and 21 of the Universal Declaration of Human Rights of December 1948 which recognized an individual's right to freedom of opinion and expression, assembly, and decisions affecting the future of one's country including elections of representatives. (UN)

Right to water – In 2002 the UN Committee on Economic, Social and Cultural Rights affirmed the right of all people to clean and sufficient quantities of fresh water as a precondition for the realization of human rights. (UN)

Rio Declaration on Environment and Development – The formal outcome of UNCED meeting in Rio de Janeiro that summarizes a consensus of principles for sustainable development.

Rio Principles – A statement of 27 principles produced by the UNCED in Rio de Janeiro in 1992 which define the rights and responsibilities of nations as they pursue human development and well-being. See Appendix I. (UNCED)

Risk – The probability of an undesirable outcome.

Risk analysis – A detailed examination performed to understand the nature of unwanted, negative consequences to human life, health, property, or the environment: to provide information regarding undesirable events; and to quantify the probabilities and expected consequences for identified risks.

River basin – A geographical area (catchment area) generally determined by the watershed limits of a water system, including surface and underground water that flow into a common terminus.

Rome Declaration on World Food Security and the World Food Summit POA (1996) – The Summit was hosted by the UN Food and Agriculture Organization (UNFAO) in Rome in November, 1996. The meeting Declaration signed by 187 heads of State and Governments reaffirmed the right of everyone to have access to safe and nutritious food, consistent with the fundamental right of everyone to be free from hunger. These officials pledged their political will and made commitments to eradicate hunger and to reduce the number of undernourished people by half, no later than 2015. The Plan of Action identifies seven major commitments, presents objectives and outlines actions that need to be taken by developed and developing countries in order to achieve these objectives.

Rovaniemi Process (AEPS) – An extensive international Arctic co-operative programme (the Arctic Environmental Protection Strategy) adopted by the eight Arctic countries: Russia, the United States, Canada, Greenland/Denmark, Iceland, Norway, Sweden, and Finland in 1991. Other countries and international organizations have observer status. Its objectives are to protect Arctic ecosystems; provide for the protection, enhancement, and restoration of natural resources; recognize the traditional and cultural needs, values, and practices of the indigenous peoples; review regularly the state of the Arctic environment; and identify, reduce, and eliminate pollution.

Rule of law – Equal protection (of human as well as property and other economic rights) and punishment under the law. The rule of law reigns over government, protecting citizens against arbitrary state action, and over society generally, governing relations among private interests. It ensures that all citizens are treated equally and are subject to the law rather than to the whims of the powerful. The rule of law is an essential precondition for accountability and predictability in both the public and private sectors. The establishment and persistence of the rule of law depend on clear communication of the rules, indiscriminate application, effective enforcement, predictable and legally enforceable methods for changing the content of laws and a citizenry that perceives the set of rules as fair, just or legitimate, and that is willing to follow it. (UNDP)

Rules of procedure – The rules that govern the organization and proceeding of a COP or other subsidiary body, including the procedures for decision-making, voting and participation. They may (and often do) vary somewhat between institutions and meetings.

S

Scenario – A plausible description of how the future may develop, based on a coherent and internally consistent set of assumptions about key relationships and driving forces. (IPCC)

Scientific and Technical Advisory Panel (STAP) – An independent advisory body to the GEF, established by UNEP in consultation with UNDP and the World Bank. It advises the GEF on the issues of global environmental management; develops and maintains a Roster of Experts; reviews selected projects; cooperates and coordinates with the scientific and technical bodies of the various MEAs; and provides a liaison between the GEF and the wider scientific and technological communities. (UNEP)

Scoping – A procedure for attempting to ensure that an environmental assessment focuses on the key environmental issues associated with a project, omitting irrelevant material. (EEA)

Second Assessment Report (SAR) – The second report of IPCC also known as "Climate Change 1995." Its conclusion (written and reviewed by some 2000 scientists) was that "the balance of evidence suggests that there is a discernible human influence on global climate" and confirmed the availability of "no-regrets" options and other const-effective strategies for combating climate change. (UNFCCC)

Secretary General – The chief administrative officer of an organisation. (MW)

Secretariat – An institution of an MEA that undertakes the day-to-day activities of coordinating the implementation of the MEA and makes arrangements for the annual meetings of the COP or periodic meetings of its subsidiary bodies.

Sector – A sociological, economic, or political subdivision of society: economically as the primary, secondary, and tertiary sectors; and socially and politically to indicate any division of human activity (or any interest group) that provides for the protection or promulgation of its interests.

Sectoral – Pertaining to a sector.

Self-contained convention – An Agreement that works using annexes and appendices and, although they may be revised periodically by the COP, requires no other protocols. (CITES)

Seoul Millennium Declaration – The declaration of the 1999 Seoul "Millennium" Conference of NGOS. The Conference was attended by more than 10,000 participants representing some 1,400 NGOs from at least 107 countries. The declaration addresses several major themes including environment, gender equality, social and economic development, education for all, and human rights. (CONGO)

Shadow pricing – A Technique sometimes used to attribute a value to a commodity or service that has no market price (or of which the market price does not reflect the true economic value) due to market failures (externalities) or market distortions.

Side event – A presentation open to all that is related to the issues being negotiated during an intergovernmental meeting.

Signator – A term that usually refers to a state that has signed an agreement or convention. (UNT)

Signature – Usually signifies that a state has consented to be bound by the treaty especially in bilateral treaties. For multilateral treaties, however, the signature is as a rule not definitive, meaning that the treaty is subject to ratification, acceptance, or approval in order to enter into force (VC).

Signature *ad referendum* – A term meaning that an agreement may be signed by a representative of a state but the agreement only becomes binding once the signature is confirmed by the responsible organ (VC)

Simple majority – Fifty percent of delegates in attendance and official proxies plus one vote.

Simultaneous translation/interpretation – Refers to the spoken word. (MW)

Sine qua non – Without which not. An indispensable condition, a prerequisite. (BLD)

Sink – Any process, activity or mechanism that removes a greenhouse gas, an aerosol or a precursor of a greenhouse gas from the atmosphere (UNFCCC).

Small Grants Programme (SGP) – A program administered by the UNDP which provides grants of up to $50,000 to grassroots groups and NGOs to demonstrate community-based strategies and technologies that address global environmental concerns. (GEF)

Social capital – Features of social organization - such as networks and values, including tolerance, inclusion, reciprocity, participation and trust - that facilitate coordination and cooperation for mutual benefit. Social capital inheres in the relations between and among actors. (UNDP)

Social issues – Issues arising due to differential access to land and other resources; conflicting demands on the same resources for conservation and economic development of traditional users; marginalization of women; religious or ethnic tensions; winners and losers in privatization and reform programs; structural exclusion of social group; poor governance, adverse social impacts, etc. (WB)

Soft law – Non-legally binding agreements. Though they may be signed by governments, they are less precise than hard law instruments and have less stringent obligation and weaker delegation. Compared to hard law instruments, they are easier to achieve, provide strategies for dealing with uncertainty, infringe less on sovereignty and facilitate compromise among differentiated actors. (Abbott and Snidal 2000)

Source – Any process or activity that releases a greenhouse gas, an aerosol or a precursor of a greenhouse gas into the atmosphere (UNFCCC).

Source Principle – The environmental damage should preferably be prevented at the source, rather than by using the "end of pipe technology". This principle also implies a preference for emission standards rather than environmental quality standards, especially to deal with water and air pollution.

South – Another name for the low- and middle-income countries as a group.

South Centre – An intergovernmental body of developing countries established on 31 July 1995 The Centre currently has 46 members and draws its strength from networking so as to carry out the designated work in cooperation with other institutions and individuals of the South. Its work is to promote solidarity, consciousness and mutual understanding among the countries and peoples of the South; promote South-South cooperation and action, South-South networking and information exchange, and developing country participation in international forums dealing with South-South and North-South matters and global concerns. It fosters convergent views and approaches among countries of the South with respect to global economic, political, and strategic issues and contributes to better mutual understanding and cooperation between the South and the North on the basis of equity and justice for all.

South-North – A term indicating collaboration or technology sharing/transfer from a "South" nation to a "North" nation.

South Pacific Regional Environment Programme (SPREP) – Headquartered in Apia, Samoa, SPREP is the Pacific Region's main environmental organization with 26 members including the U.S., Australia, New Zealand and France. Its mandate is to promote cooperation in the South Pacific Region, provide assistance in order to protect and improve its environment and ensure sustainable development for present and future generations.

South-South – A term indicating collaboration or technology sharing between low income and /or middle income developing countries. (UNT)

Sovereignty – This difficult concept has both emotional and rational components and is often a matter of perception. It relates to the state's monopoly on the use of force and its position as final authority over matters within its territory. State sovereignty is diminished with the rise of international corporations, multinational organizations and transnational forces like religion, ethnicity, and culture.

Specific funds – Monies donated for a specific purpose but within the overall mandate of an organization but normally in addition to regular funds.

Square brackets [ ] – Symbols used during treaty negotiations to indicate that a section of text is being discussed but has not yet been agreed upon.

Stability – Unchanging with time. This can be a static state (nothing changes) or a steady state (resource flows occur). (CSG)

Stakeholder – An institution, organization, or group that has some interest in a particular sector or system or outcome of a project, program or policy initiative. (EEA)

State – The set of political institutions whose specific concern is with the social and political organization and management, in the name of the common interest, within a determined territory. (UNDP)

State of the World Forum (SWF) – Formed in 1995, SWF is a non profit organization dedicated to "establishing a global network of leaders, citizens and institutions committed to discerning and implementing the principles, values and actions necessary to guide humanity wisely as it gives shape to an increasingly global and interdependent civilization." (SWF)

Statement of Forest Principles (UNCED) – The "Non-legally Binding Authoritative Statement of Principles for a Global Consensus on the Management, Conservation and Sustainable Development of all Types of Forests" was agreed to at UNCED in 1992 and addresses issues related to sustainable forest management. This was the first global consensus on forests which attempted to deal with people who want to protect forests for environmental and cultural reasons together with those whose who depend on trees and other forest life for economic development. The hope is that these principles will form the basis of further negotiations toward a binding agreement.

Stockholm Convention – An international legally binding instrument (opened for signature on 22 May 2001) for implementing international action on twelve persistent organic pollutants (POPs) grouped into three categories: 1) Pesticides: aldrin, chlordane, DDT, dieldrin, endrin, heptachlor, mirex and toxaphene; 2) industrial chemicals: hexachlorobenzene (HCB) and polychlorinated biphenyls (PCBs); and 3) unintended by-products: dioxins and furans. The Convention will come into force when 50 Parties have ratified.

Stockholm Declaration on the Human Environment – The Declaration produced at the United Nations Conference on the Human Environment that met at Stockholm from 5-16 June 1972 which supports a common outlook and principles to inspire and guide the peoples of the world in the preservation and enhancement of the human environment. See Appendix I.

Straddling stock – A population of organisms that travels between the exclusive economic zones of two or more countries, or between them and the high seas. (BCHM)

Strange attractors – Areas around which behaviors and interactions tend to occur.

Strategic environmental assessment (SEA) – A technique similar to environmental impact assessment (EIA) but normally applied to policies, plans, program and groups of projects. There are two main types of SEA: Sectoral, which are applied when many new projects fall within one sector and Regional which are applied to cover development within a region. (EEA)

Strategy – A framework guiding those choices that determine the nature and direction to attain the objective.

Structural adjustment loans – Large loans made by the World Bank or IMF to developing countries that may carry strict financial and budgetary obligations or required reforms intended to open recipient countries to private investment and increase the recipient's competitiveness in the global economy. Reforms are usually orientated towards liberalization, privatization, and reduction in government expenditure.

Subsidiarity, principle of – A hierarchical form of government where decision making is made at the level of authority most responsive to the public. (UNCHS). A principle that recognizes that actions will occur at different levels of jurisdiction depending on the nature of the issue. It assigns priority to the lowest jurisdictional level consistent with effectiveness. (IISD)

Subsidiary Body for Implementation – Assists the Conference of the Parties (COP) in the assessment and review of the effective implementation of a convention. (UNFCCC)

Subsidiary Body on Scientific, Technical and Technological Advice (SBSTTA) – The SBSTTA is a body of the Conference of the Parties (COP) that reports regularly on its assessments of the status the subject matter of the Convention; the measures taken by the Parties in accordance with the provisions of the Convention; and, it responds to questions put to it by the COP. (CBD; UNFCCC)

Subsidy – A direct or indirect benefit granted by a government for the production or distribution (including export) of a good.

Succession – The more or less predictable changes in the species composition on a given site.

Sui generis – A Latin term meaning unique or peculiar, in a class of its own.

Summits of the Americas Process – There have been three formal Summits of the Americas (Miami, USA December 1994; Santiago, Chile April 1998; and Quebec City, Canada April 2002) and one specialized Summit of the Americas on sustainable development (Santa Cruz, Bolivia December 1996). Selected principles from the Santa Cruz Summit are included here. Subsequent formal Summits have endorsed the principles elaborated in Santa Cruz. The Americas is the only hemisphere that has convened a Presidential-level Summit on Sustainable Development that attempted to prioritize the initiatives included in *Agenda 21*. (UNCED)

Supplementarity – A term referring to whether or not the Parties of The Kyoto Protocol, which use flexibility mechanisms such as emissions trading to lower greenhouse gas mitigation costs, also institute adequate domestic energy and other policies to achieve long-term greenhouse gas reduction goals.

Sustainability – See "Clarity and Chaos" in the essay.

Sustainability indicator – A variable, a pointer, an index of a complex phenomenon. Its fluctuations reveal the variations in components of the ecosystem, the resource or the sector. The position and trend of the indicator in relation to the criteria indicate the present state and dynamics of the system. Ideally, composite indicators are needed, the position and trajectory of which, within a system of reference of related criteria, would allow simple holistic assessment of sustainability. One can distinguish indicators of the state of the system, pressure (or stress, driving forces) on the system, and response (reflecting action taken to mitigate, reduce, eliminate or compensate for the stress). (EEA)

Sustainability, principle of – A principle which implies a relationship between the rate and type of use of natural resources and the ability to support a defined lifestyle. (RFF)

Sustainable Community - Fosters healthy people and biodiversity, and the linkages between them; continuously invests in capacities, institutions and partnerships that protect, restore, and enhance natural, social and economic capital; actively monitors and disseminates knowledge and know-how; promotes inclusive, collaborative, stakeholder-driven planning and adaptive action in response to changing conditions; prefers local products and services that result in triple bottom-line profits – social, economic and environmental; and has increasing options and opportunities because its citizens choose to internalize the local, regional, global, and inter-generational impacts of their decisions and actions. (SNW)

Sustainable Consumption – The consumption of services and products that respond to basic needs and bring a better quality of life, while minimizing use of natural resources and toxic materials as well as the generation of wastes and pollutants over the life-cycle of a product or service, so as not to jeopardize the needs of future generations. (UNCSD)

Sustainable development – See "Clarity and Chaos" in the essay, especially footnote 76.

Sustainable Enterprise - Understands that the production and sale of goods and services must incorporate the role of and impact on nature and society; pursues triple bottom line profits—economic, social and environmental; practices eco-efficiency and commits to moving toward eco-effectiveness (i.e. where the by-products of production are inputs for other processes – zero waste); builds strong customer base and expands market share by helping consumers consume less, serving their true needs; and, engenders loyal shareholders and employees. (SNW)

Sustainable forest management – Forest management activities that benefit people and the resource base today, but do not compromise the resource for future generations. (USFS). The stewardship and use of forests and forest lands in a way, and at a rate, that maintains their biodiversity, productivity, regeneration capacity, vitality and their potential to fulfill, now and in the future, relevant ecological, economic and social functions, at local, national, and global levels, and that does not cause damage to other ecosystems. (UNFF)

Sustainable human development (SHD) – A paradigm of development that puts people and their ongoing needs and aspirations at the center of its concerns; that not only generates economic growth but distributes its benefits equitably, that regenerates the environment, that empowers people and that, because of their existing position of fewer opportunities, gives priority to the poor, enlarging their choices and opportunities and providing for their participation in decisions that affect their lives. UNDP's core mission, focused on building human and institutional capacity for poverty elimination, livelihood and job creation, the advancement of women and environmental protection. SHD differs from Human Development concepts by more attention towards the factor of time. (UNDP)

Sustainable livelihood – A livelihood is sustainable when it can cope with and recover from stresses and shocks while maintaining or enhancing its capabilities and assets.

Sustainable use (of biodiversity) – The use of components of biological diversity in a way and at a rate that does not lead to the long-term decline of biological diversity, thereby maintaining its potential to meet the needs and aspirations of present and future generations. (CBD)

Sustained yield – The yield that a renewable resource can produce continuously at a given intensity of management.

Symbols – Symbols, including units of measurement (UM) used in United Nations documents.

- b billion
- © copyright
- € euro
- g giga
- ha hectare (2.47 acres)
- k kilo
- l litre, liter

- £ British pound
- m metre, meter
- m^2 square metre, square meter
- m^3 cubic metre, cubic meter
- masl meters above sea level
- M mega
- M million
- [ ] square brackets
- ® registered
- s second
- T tera
- t ton
- ™ trademark
- US$ United States dollar
- ¥ Japanese yen

Synergy – The result of a combined action or operation of two or more parties, systems, or parts, in which the results are greater than they would be acting separately.

T

Table – A term which, depending on who is using it, can mean to either take up or halt debate on an issue. To place on the agenda or to remove (as a parliamentary motion) from consideration indefinitely.

Tacit Knowledge - The accumulated portion of experiences and skills within people gained through cognition in one's lifetime. (Nonaka and Takeuchi)

Take-back Effect – See "**Offsetting Effect**."

Target groups – The main stakeholders of a project that are expected to gain from the results of that project; sectors of the population that a project aims to reach in order to address their needs based on gender considerations and their socio-economic characteristics.

Technology transfer – The transfer of knowledge or equipment to enable the manufacture of a product, the application of a process, or the rendering of a service. (BCHM)

Third State – A state not a party to the treaty. (VC)

Third World – A term referring to the areas of the world that are the least developed that consists of Latin America, Africa, and South and East Asia.

Threatened species (1) – Species that are genetically impoverished, of low fecundity, dependent on patchy or unpredictable resources, extremely variable in population density, persecuted or otherwise prone to extinction in human dominated landscapes. (GBA)

Threatened species (2) – A classification referring to a species that is likely to become endangered within the near future, throughout all or a significant portion of its range, vulnerable taxa. (IUCN)

Three pillars of sustainable development – A commonly accepted idea that sustainable development is based on the sustainability of its three components: economical sustainability, social sustainability, and "ecological" or "environmental" sustainability.

Top-up funds – Funds that are in addition to those requested for a project for the purposes of enabling the project to move forward.

Top-up projects – Enabling projects that are different from, but necessary for, the success of projects designed to meet a specific goal.

Tradable emissions – An approach to limiting the global discharge of greenhouse emissions by allowing polluters to "trade" the costs of emissions for proactive conservation initiatives in related sectors. An example would be carbon emissions from a power station "traded" for support to carbon sequestration in forestry.

Traditional resource rights (TRR) – Similar to intellectual property rights, but the terms denotes broader "bundles of rights" that include human rights, land rights, religious rights, and cultural property. (BCHM)

TRAFFIC – A wildlife trade monitoring group sponsored by IUCN and WWF.

Tranche – A work program that contains a date, project component and funding level. (GEF)

Transboundary – Describes collaboration activities or any system or externality extending beyond one region or nation state at any scale (cross-border, inter-regional, transnational) and is thus a generic term for collaboration extending over national borders.

Transboundary diagnostic analysis (TDA) – A study undertaken to identify and evaluate the major environmental problems and their root societal and economic causes responsible for the deterioration of international waters. Further, the TDA is a mechanism to define priority actions and co-operative efforts between GEF and countries-participants. The study includes natural and social scientists and economists as well as representatives of the area's stakeholders. (GEF-IW)

Transboundary governance – A term referring to either a formal or informal framework to manage a natural resource or to address an environmental problem shared between two or more countries. (OAS)

Transboundary pollution – Polluted air and water, or any other contaminated waste that is generated in one country and transmitted to others. (EEA)

Transition countries – A term referring to a class of 24 countries from the former Soviet Union (Commenwealth of Independent States) and the former Warsaw Pact.

Translation – A term referring to translating the written or spoken word into other languages.

Transparency – A policy of sharing information and acting in such a way so as to allow stakeholders to gather information that may be critical to uncovering abuses and defending their interests. Transparent systems have clear procedures for public decision-making and open channels of communication between stakeholders and officials, and make a wide range of information accessible. (UNDP)

Transparency International – Formed in 1993 and based in Berlin, Transparency International is a non-governmental organization dedicated to increasing government accountability and curbing both international and national corruption. It works to ensure that the agendas of international organizations give high priority to curbing corruption. Both the TI Secretariat and TI National Chapters around the world actively monitor the implementation of the anti-corruption instruments of the OECD, the Council of Europe, the European Union and the Organization of American States. (TI)

Treaty – An international agreement concluded between States in written form and governed by international law, whether embodied in a single instrument or in two or more related instruments and whatever its particular designation. (VC)

Tripartite Agreement – An October 28, 1991 agreement formally signed by the heads of the World Bank, UNDP and UNEP to establish and operate the GEF.

Triple-bottom line – The whole set of values, issues and processes companies must address in order to minimize the harm resulting from their activities and to create economic, social and environmental value. The three bottom lines represent society, the economy and the environment. Society depends on the economy and the economy depends on the global ecosystem, whose health is the ultimate bottom line. (UNEP)

Trust Fund for the Convention on Biological Diversity – An institution established by the COP-CBD to administer funds for the administration of the convention including the functions of the Secretariat. The Trust Fund is financed from contributions made by Parties to the Conventions based on a set scale, additional contributions by the Parties, contributions from States not Parties to the Convention and governmental, intergovernmental and non-governmental organizations. (CBD)

Trust funds – Extra-budgetary programme resources that help meet the objectives of UNEP that are negotiated between UNEP and the donor (or several donors) and form separate accounts. (UNEP)

Twinning arrangements – Arrangements made to facilitate sharing of knowledge and technology transfer between institutions having similar mandates in different countries.

Type 1 Outcome – A tangible outcome of a conference, generally a treaty.

Type 2 Outcome – A tangible outcome of a conference, generally partnerships.

U

Umbrella convention – A stand-alone convention that can foster other conventions (primarily regional conventions) that helps to meet its objectives. (CMS)

Umbrella Group – A coalition of like-minded governments established for purposes of voting within the UNFCCC (Japan, U.S.A., Iceland, Canada, Australia, Norway, New Zealand, Russian Federation, and Ukraine).

Unanimity – A decision having the agreement and consent of all participants.

UN Water – An inter-agency structure of all 24 agencies in the UN system with activities related to water in all its aspects. This group is responsible for steering the World Water Assessment Program and its principle output, the World Water Assessment Report, and serves as an action forum to provide a multi-stakeholder mechanism for engagement with extra UN bodies through the World Water Implementation Programme.

Uncertainty – Distinguished from risk by the fact that there is no scientific basis upon which to formulate any probabilistic argument.

UNESCO Centre University of Ulster – A university research centre that concentrates on the themes of education for pluralism, human rights and democracy in local, national and international contexts.

UNESCO - IHE Institute for Water Education – Located in Delft, The Netherlands, UNESCO-IHE is a post-graduate training institute in water sciences, engineering and policy that aims to contribute to the education and training of professionals and to build the capacity of sector organizations, knowledge centers and other institutions active in the fields of water, environment, and infrastructure in developing countries and countries in transition. (UNESCO)

UNESCO International Hydrological Programme (IHP) – UNESCO's intergovernmental scientific cooperative programme in water resources is a vehicle through which Member States can upgrade their knowledge of the water cycle and thereby increase their capacity to better manage and develop their water resources. IHP aims at the improvement of the scientific and technological basis for the development of methods for the rational management of water resources, including the protection of the environment. (UNESCO)

UNESCO – IHE Partnership for Water Education and Research (PoWER) - A network of seventeen autonomous collaborating centers in all regions of the developing world that aims to conduct and disseminate research into all aspects of integrated water resource management and capacity building. (UNESCO)

UNESCO "Water Family"– An informal union of the Division of Water Sciences of UNESCO Headquarters, UNESCO-IHP (centers, chairs, networks and centers) and UNESCO-IHE that collaborate on water related issues aimed at achieving the goals of the UN Decade for Action: "Water for Life" and the UN Millennium Development Goals, launched in July, 2003 in Delft, The Netherlands. (UNESCO)

UNese – A slang term referring to the array of jargon and acronyms used by UN insiders, making full understanding by outsiders very difficult.

UNGASS (Earth Summit +5) – The 19th special session of the UNGA convened in June of 1997 to review the outstanding issues not fully resolved at the UNCED including finance, climate change, freshwater management, and forest management. This session adopted the Programme for the Further Implementation of *Agenda 21* (A/RES/S-19/2). It assessed progress since UNCED, examined implementation, and established the CSD's work programme for the period 1998-2002, leading up to WSSD.

United Nations – The United Nations Conference on International Organization opened in San Francisco on April 25, 1945 where delegates of 50 nations discussed and modified the original Dumbarton Oaks proposals. On June 26 the United Nations Charter was completed and signed. By Oct. 24, 1945, the required number of nations had ratified the charter and the United Nations came officially into existence. Trygve Lie, foreign minister of Norway, was elected the first secretary-general.

United Nations Capital Development Fund (UNCDF) – The UNCDF was established by the UNGA in 1966 as a semi-autonomous unit within UNDP that develops new solutions for poverty reduction in least developed countries.

United Nations Centre for Human Settlements (UNCHS) – UNCHS (Habitat), located in Nairobi, Kenya, was established in 1978, two years after the United Nations Conference on Human Settlements held in Vancouver, Canada. The Centre serves as the lead agency for United Nation's human settlement development activities, as well as for the global exchange of information about human settlements, conditions and trends.

United Nations Charter – The UN Charter is the founding document of the United Nations, which was opened for signature in June 1945 and entered into force in October 1945.

United Nations Conference on Environment and Development (UNCED) – The United Nations Conference on Environment and Development was held from June 3 through June 14, 1992, in Rio de Janeiro, Brazil. Five major agreements on global environmental issues were signed. Two of these, The Framework Convention on Climate Change and The Convention on Biological Diversity are formal treaties whose provisions may eventually be binding on the parties. The other three UNCED agreements are non-binding statements on the relationship between sustainable environmental practices and the pursuit of social and socioeconomic development: *Agenda 21* is a wide-ranging assessment of social and economic sectors with goals for improving environmental and developmental impact of each; the *Rio Declaration* summarizes consensus principles of sustainable development, and the *Statement on Forest Principles* pledges parties to more sustainable use of forest resources.

United Nations Convention on the Law of the Sea (UNCLOS) – UNCLOS provides a global framework for the protection and management of the marine environment and the conservation and management of its living and non-living resources. It became operational in November 1994.

United Nations Convention to Combat Desertification in Countries Experiencing Serious Drought and/or Desertification, Particularly in Africa (1994) – As the question of how to tackle desertification was still a major concern at the time of UNCED, *Agenda 21* supported a new, integrated approach to the problem emphasizing action to promote sustainable development at the community level. It also called on the UNGA to establish an Intergovernmental Negotiating Committee to prepare, by June 1994, a Convention to Combat Desertification. In December 1992, the General Assembly agreed by adopting Resolution 47/188. The Committee completed its negotiations in five sessions and the Convention was adopted in Paris on 17 June 1994. It entered into force on 26 December 1996.

United Nations Decade – A ten year period during wich a particular theme is given priority in terms of funding (regular and extrabudgetary) such as the UN Decade for Sustainable Education (2005-2015); The UN Decade: "Water for Life" (2005-2015). (UN)

United Nations Development Group – A consortium of the heads of all United Nations development funds, programs and departments headed by the Administrator of the United Nations Development Programme.

United Nations Economic Commission for Europe (UNECE) – The UNECE was instituted in 1947 as a forum at which the countries of North America, Central and Eastern Europe and Central Asia come together to strengthen economic co-operation. UNECE focuses on economic analysis, environment and human settlements, statistics, sustainable energy, trade, industry and enterprise development, timber and transport.

United Nations Economic Commission for Latin America and the Caribbean (ECLAC) – ECLAC was established in February 1948 as a regional agency of the UN to cooperate with the governments of Latin America in the economic and social development of the region; the scope of its activities was subsequently expanded to include the countries of the Caribbean. ECLAC carries out studies and programs in collaboration with the governments of the region in integration, foreign trade, agricultural production, industrial development, transport and communications, statistics, natural resources, the environment, science and technology, and transnational corporations. (ECLAC)

United Nations Educational, Scientific, and Cultural Organization (UNESCO) – UNESCO was founded on 16th of November 1945. Its headquarters are in Paris, France with 73 field offices and units in different parts of the world. The main objective of UNESCO is to contribute to peace and security in the world by promoting collaboration among nations through education, science, culture, and communication in order to further universal respect for justice, for the rule of law and for the human rights and fundamental freedoms.

United Nations Environment Programme (UNEP) – UNEP was created as a consequence of the 1972 Stockholm Conference on the Human Environment. Its mission is to be the leading environmental authority within the UN system that sets a global environmental agenda; promotes the coherent implementation of the environmental dimension of sustainable development; serves as an authoritative advocate for the global environment; and encourages international cooperation and action, based on the best scientific and technical capabilities available.

United Nations Food and Agriculture Organization (FAO) – FAO was founded in 1945 with a mandate to raise levels of nutrition and standards of living, to improve agricultural productivity, and to better the condition of rural populations. It is one of the largest specialized agencies in the United Nations system and the lead agency for agriculture, forestry, fisheries and rural development. An intergovernmental organization, FAO has 183 member countries plus one member organization, the European Community. (FAO)

United Nations Forum on Forests (UNFF) – The UNFF is subsidiary body of ECOSOC established to promote the management, conservation and sustainable development of all types of forests. Its purpose is to: promote the implementation of internationally agreed actions on forests at the national, regional and global levels; provide a coherent, transparent and participatory global framework for policy implementation, coordination and development; and carry out principal functions, based on the Rio Declaration, the Forest Principles, Chapter 11 of *Agenda 21*, and the outcomes of the IPF and the IFF.

United Nations Framework Convention on Climate Change (UNFCCC) – The UNFCCC was signed in 1992 at the Earth Summit in Rio de Janeiro by more than 150 countries. It consists of a Preamble, 26 Articles and two Annexes and has as its objective to achieve the stabilization of greenhouse gas concentrations in the atmosphere at a level that would prevent dangerous anthropogenic interference with the climate system...within a timeframe sufficient to allow ecosystems to adapt naturally to climate change, to ensure that food production is not threatened and to enable economic development to proceed in a sustainable manner. The treaty took effect in March 1994 upon the ratification of more than 50 countries; some 186 nations have now ratified. (UNFCCC)

U

United Nations High Commissioner for Refugees (UNHCR) – The UNHCR was established in 1951 and is mandated to lead and coordinate international action for the worldwide protection of refugees and the resolution of refugee problems. Its primary purpose is to safeguard the rights and well-being of refugees and ensure that everyone can: exercise the right to seek asylum; find safe refuge in another state; and, to return home voluntarily.

Universal Declaration of Human Rights – A declaration made by the United Nations General Assembly in 1948 which outlines a common standard for all people and all nations regarding the human rights and freedoms (See Appendix 5).

UNspeak – See **UNese**.

Uruguay Round – The round of multilateral trade negotiations under the GATT that commenced in 1986 and was completed in 1994 with the creation of the WTO. (DD)

User-pays Principle – A variation of the polluter-pays principle that calls upon the user of a natural resource to bear the cost of running down natural capital.

V

Valuation – The process of determining the current worth of an asset or the estimated worth of a thing or process. It differs from price, which does not always afford a true criterion of value. (LLL)

Variety – Plant grouping, within a single botanical taxon of the lowest known rank, defined by the reproducible expression of its distinguishing and other genetic characteristics (CBD)

Vienna Convention of 1963 – The Vienna Convention on Civil Liability for Nuclear Damage (1963). This Convention preceded the adoption of the Protocol to Amend the 1963 Vienna Convention on Civil Liability for Nuclear Damage and the Convention on Supplementary Compensation for Nuclear Damage adopted in 1997.

Vienna Convention of 1969 – The Vienna Convention on the Law of Treaties was signed at Vienna, May 23, 1969 and entered into Force, 27 January 1980. It is the source for definitions and principles that guide international treaties. Previously, treaty rules were based on customary international law or were a part of the general principles of law. The Vienna Convention codified these rules and now governs all the international treaties since it entered into force. Although the Vienna Convention does not apply to treaties concluded before its entry into force, it is *de facto* applied since it incorporates the customary rules that were applicable before this date.

Vienna Convention of 1985 – The Vienna Convention for the Protection of the Ozone Layer is a convention that gives priority to research, cooperation among countries, and exchange of information regarding substances and activities which may harm the ozone layer.

Vienna Convention of 1986 – The Vienna Convention on the Law of Treaties between States and International Organizations or between International Organizations (1986) reflects international customary law but it has yet to enter into force.

Vienna Convention on Civil Liability for Nuclear Damage – A UN Convention on liability and compensation for nuclear damage similar to the

Paris Convention. It was recently revised by means of the 1997 Protocol and a new Convention on Supplementary Compensation was agreed to provide additional compensation, although this has not yet entered into force. The Vienna Convention and Paris Convention are linked by the Joint Protocol relating to the Application of the Vienna Convention and the Paris Convention.

Vienna setting – A negotiating format modeled after the final negotiations on the Cartagena Protocol on Biosafety that involves spokespersons from the major negotiating groups.

Virtual water/issues – See **Water**.

Voluntary agreement – A set of policy instruments introduced in several countries to prevent industrial pollution. Pollution reduction is reached through a contract between government and industry and may include regulation and taxation.

Voluntary contribution – Funds donated by a state in excess of its assessed contribution that may be based on a separate appeal or for a designated purpose. (UN)

Voluntary fund – A fund created for a specific purpose and to which countries are encouraged to contribute. (UN)

Vulnerability – The propensity of a population group to experience substantial damage, disruption and casualties as a result of a hazard. (DFID) A condition characterized by higher risk and reduced ability to cope with shock or negative impacts. It may be based on socio-economic condition, gender, age, disability, ethnicity, or other criteria that influence people's ability to access resources and development opportunities. Vulnerability is always contextual, and must be assessed in the context of a specific situation and time. Good practice in World Bank projects indicates that development interventions and support mechanisms should assess vulnerability, and target interventions to be appropriate and reduce risk for those deemed as vulnerable. (WB)

Vulnerable – An IUCN category of taxa believed likely to move into the Endangered category in the near future if the causal factors continue operating. Included are taxa of which most or all the populations are decreasing because of over-exploitation, extensive habitat or other environmental disturbance; taxa with populations that have been seriously depleted and whose ultimate security is not yet assured; and taxa with populations that are still abundant but are under threat from serious adverse factors throughout their range. (IUCN)

Vulnerable taxa – A classification referring to a species that is likely to become endangered within the foreseeable future, throughout all or a significant portion of its range. (IUCN)

W

Wadi Hydrology – A project initiated by UNESCO, through its International Hydrology Programme on "Wadi Hydrology," that focuses on the sustainable management of ephemeral rivers. The Arabic term "Wadi" refers to a seasonal water source in the world's arid regions which are an excellent and at times the only source of freshwater. (UNESCO)

Washington Consensus – A term describing a set of policy reforms which most official Washington financial institutions thought would be good for Latin America. The reforms include fiscal discipline, a redirection of public expenditure priorities toward primary health care, primary education, infrastructure, tax reform, interest rate liberalization, a competitive exchange rate, trade liberalization, privatization, deregulation, and secure property rights. However, there are now other, often mutually exclusive, definitions of the term that complicate its use.

Water

- **Blue water/issues** – A term relating to surface and groundwater freshwater resources. (UN)

- **Black water/issues** – Refers to water that contains or is contaminated with faecal material, (sanitary engineers); fossil groundwater (hydrologists, geologists), water from swamps or other wetlands (wetland ecologists). (UN)

- **Brown water/issues** - A term referring to highly acidic waters normally produced in swamps, wetlands, or wet environments with a low flow rate or near-stagnate condition (wetland ecologists); and to raw wastewater (sanitary engineers).

- **Gray water/issues** – A term referring to wastewater from shower, laundry and other (household and industrial) cleaning activities which does not contain faecal material (sanitary engineers); and to treated wastewater for reuse (hydrologists). (UNFAO)

- **Green water/issues** – Water in the biosphere or water that supports all non irrigated vegetation, including forests and woodlands, grasslands and rain-fed crops or the water available in the root zone (UNFAO); all issues related to "water for nature." (UN)

- **Virtual Water** – A term referring to water used for food production or for producing industrial products. This term has also been used to reflect the potential for producing new products that could be realized if more water was available or new investments were made to improve water resource management. (UN)

- **White water** – A term referring to water lost through evaporation, through interception or directly from the soil. (UNFAO); and to the oceans, rapids (recreationists). (USGS)

Water Cooperation Facility (WCF) - A UNESCO / World Water Council (WWC) initiative linking these two bodies together with other pivotal organizations, the Permanent Court of Arbitration, and the University Partnership for Transboundary Waters (UPTW) aimed at resolving transboundary water disputes and serving as a nexus for education and technical expertise in dispute resolution techniques and case studies.

Watershed – The divide separating one drainage basin from another. However, although "drainage basin" is preferred, over the years, use of the term "watershed" to signify a drainage basin or catchment area has come to predominate. Drainage divide, or just divide, is used to denote the boundary between one drainage area and another. Used alone, the term "watershed" is ambiguous and should not be used unless the intended meaning is made clear. (USGS)

WEHAB (Water, Energy, Health, Agriculture and Biodiversity) objectives articulated at WSSD including: water supply and sanitation; cleaner production and recycling; energy efficiency and conservation, renewable energy and clean coal technology; emergencies and disaster preparedness and response, including urban security; post shock and conflict restoration, rehabilitation and reconstruction; engaging engineers in decision making, policy making and planning.

Weighted Majority – A majority where additional weight is given to some countries for varied specific reasons such as monetary contributions.

Westphalia Settlement of 1648 – The treaty that spelled out the conditions to end the Thirty Year War that is generally considered to be the origin of modern state order. This treaty recognized the absolute sovereignty and legal equality of states as the basis for international order.

W-E-T - A UNESCO-IHP Phase VI program which provides a coherent approach in the area of partnerships in water-related capacity building. Specifically it address four priority areas: unity of education and research in higher education, facilitation of high intensity networks, quality assurance and assessment, and raising public awareness as an obligation of water professionals and educators.

Wetlands – Areas of marsh, fen, peatland or water whether natural or artificial, permanent or temporary, with water that is static or flowing, fresh, brackish or salt, including areas of marine water the depth of which at low tide does not exceed six metres (See **Ramsar Convention**).

White water/issues – See **Water**.

White water to blue water - A U.S. government initiative launched at the WSSD in Johannesburg to promote integrated river basin management in the Caribbean.

WHO Health Policy for All in the 21st Century – A global health policy aimed at meeting the major challenges in health during the next decades. The policy was developed in consultation with all WHO national and international partners and evolved from the Health-For-All goal subscribed to in 1979.

Willingness to pay – The amount an individual is willing to pay to acquire some good or service. This may be elicited from stated or revealed preference approaches. (EEA)

Win-win options – In the context of multi-criteria analysis those options which have the highest scores on both selected criteria. Also more broadly used to characterize options with potential positive welfare implications for both parties involved.

Withdrawal – Termination of a party to a treaty, usually with twelve month's notice for multilateral treaties and automatically after a fixed period or when the projects they deal with are completed for some bilateral treaties. A party may withdraw from or denounce a treaty in accordance with the withdrawal or denunciation provisions of the treaty (article 54(a) of the 1969 Vienna Convention) or, with the consent of all parties to a convention (article 54(b) of the 1969 Vienna Convention). In the case of a treaty that is silent on withdrawal or denunciation, withdrawal is possible by giving at least 12 months' notice provided that it is established that the parties intended to admit the possibility of denunciation or withdrawal; or a right of denunciation or withdrawal may be implied by the nature of the treaty. (VC)

Working Group – A group convened by a COP or one of the subsidiary bodies to work on large-scale issues. Its head, or chair, of the working group is designated by the president of the body that created it and its membership is open to all Parties to the MEA or subsidiary body. (JG)

World Bank – Established in 1944 to help Europe recover from the devastations of World War II, the World Bank is also known as the International Bank for Reconstruction and Development (IBRD). It now consists of a group of five institutions (IBRD, IDA, IFC, MIGA, ICSID) that are the world's largest source of development assistance (over $15 billion in loans in FY 2000) to its 100 plus client countries.

World Business Council for Sustainable Development (WBCSD) – Formed in 1991, the World Business Council for Sustainable Development (WBCSD) is a coalition of 150 international companies (2002) united by a shared commitment to sustainable development via the three pillars of economic growth, ecological balance and social progress. Members are drawn from more than 30 countries and 20 major industrial sectors. Its mission is to provide business leadership as a catalyst for change toward sustainable development, and to promote the role of eco-efficiency, innovation and corporate social responsibility. (WBCSD)

World Charter for Nature – Ethical guidelines for the relationships between humans and the rest of nature that were accepted by the world community as UNGA resolution 37/7 (1982).

World Commission on the Social Dimension of Globalization – A commission of the ILO created in November of 2001 that is mandated to analyze the social dimensions of globalization and make proposals to correct any imbalances. The commission is made up of 26 internationally eminent persons. Reports its findings and recommendations can be found at http://www.ilo.org/public/english/wcsdg/commission.htm.

World Conservation Monitoring Centre (WCMC) – The World Conservation Monitoring Centre of UNEP provides information services on conservation and sustainable use of the world's living resources, and helps others to develop information systems of their own. It was established in 2000 as the world biodiversity information and assessment center but has roots that go back to 1979 when IUCN established a Cambridge office to monitor endangered species. In 1988 the independent, non-profit World Conservation Monitoring Centre was founded jointly by IUCN, WWF and UNEP. (WCMC)

World Conservation Strategy (WCS) – The WCS is a report and long-range worldwide conservation plan sponsored by the IUCN, UNEP, and WWF. Published in 1980, it was one of the first written and widely distributed calls for sustainable development.

World Conservation Union (IUCN) – The World Conservation Union was founded in 1948 and brings together 78 states, 112 government agencies, 735 NGOs, 35 affiliates, and some 10,000 scientists and experts from 181 countries in a unique worldwide partnership. Its mission is to influence, encourage, and assist societies throughout the world to conserve the integrity and diversity of nature and to ensure that any use of natural resources is equitable and ecologically sustainable. (IUCN)

World Declaration on Nutrition and the Plan of Action for Nutrition (1992) – The outputs of the 1992 UNFAO-sponsored meeting held in Rome to prepare for the World Food Summit.

World Economic Forum ("Davos Symposium") – The World Economic Forum, based in Switzerland, is an independent organization committed to improving the state of the world. Funded by the contributions of 1,000 of the world's foremost corporations, the Forum acts in the spirit of entrepreneurship in the global public interest to further economic growth and social progress. The Forum serves its members and society by creating partnerships between and among business, political, intellectual and other leaders of society to define, discuss and advance key issues on the global agenda. Incorporated in 1971 as a foundation, the World Economic Forum is impartial and not-for-profit, and is tied to no political, partisan, or national interests. In 1995, the Forum was awarded NGO consultative status with the Economic and Social Council of the United Nations. (WEF)

World Health Organization (WHO) – An international organization established in 1948 with the goal of improving human health. WHO assists countries in strengthening their health services, provides technical assistance in health emergencies, promotes disease prevention and control, and promulgates international food safety and medical standards. It currently has 191 member countries. (WHO)

World Heritage Site – Cultural and natural properties, submitted by the States Parties to The Convention Concerning the Protection of the World Cultural and Natural Heritage that are considered to be of outstanding universal value.

World Resources Institute (WRI) – The World Resources Institute (WRI) is a policy research center created in late 1982 to help governments, international organizations, and private businesses understand how to meet basic human needs and nurture economic growth without undermining the natural resources and environmental integrity on which life, economic vitality, and international security depend. The Institute's current areas of policy research include forests, biological diversity, sustainable agriculture, energy, climate change, atmospheric pollution, economic incentives for sustainable development, and resource and environmental information. (WRI)

World Social Forum (WSF) – The World Social Forum is an open meeting place for reflective thinking, democratic debate of ideas, formulation of proposals, free exchange of experiences and interlinking for effective action, by groups and movements of civil society opposed to neoliberalism, and domination of the world by capital and any form of imperialism. The WSF wishes to build a planetary society centered on the human person. (WSF)

World Summit on Sustainable Development (WSSD) – A world meeting held from 26 August - 4 September 2002, at the Sandton Convention

Centre in Johannesburg, South Africa. According to UN General Assembly (UNGA) Resolution 55/199, the meeting's goal was to hold a ten-year review of the 1992 UN Conference on Environment and Development (UNCED) at the Summit level to reinvigorate global commitment to sustainable development. Over 21,000 participants from 191 countries attended and included representatives from intergovernmental and non-governmental organizations, the private sector, civil society, academia and the scientific community. The WSSD adopted two main documents: the Plan of Implementation and the Johannesburg Declaration on Sustainable Development. (WSSD)

World Trade Organization (WTO) – An organization based in Geneva set up in 1995 to implement and enforce the Uruguay Round Agreement. Replaces the General Agreement on Tariffs and Trade (GATT). WTO establishes the legal and institutional foundations of the international trading system. It determines government obligations in trade legislation and regulation, and specifies trade dispute resolution mechanisms. It currently has 142 member countries. (WTO)

Worldwide Fund for Nature (WWF) – Formerly known as the World Wildlife Fund, the WWF is an independent conservation organization with almost five million supporters from five continents. Since 1985 its 28 National Organizations, 24 Programme Offices and 4 Associates have invested over US$1,165 million in more than 11,000 projects in 130 countries. (WWF)

WSSD Plan of Implementation – A framework for action to implement the commitments agreed at UNCED. The plan of implementation has eleven chapters: introduction; poverty eradication; consumption and production; the natural resource base; health; small island developing States; Africa; other regional initiatives; means of implementation; and institutional framework. (See **World Summit on Sustainable Development**).

Z

Zero sum – A term originating in game theory where the sum of wins and losses in a game always equals zero for each set of strategies chosen. In governance, it means that if one party wins, the other party looses.

Zero Waste Alliance (ZWA) – An initiative of the International Sustainable Development Foundation aimed at introducing life cycle responsibility and green chemistry principles into manufacturing processes.

Zoning - The arrangement or partitioning of land areas for various types and intensities of usage in natural or managed protected lands, cities, boroughs, townships or other political jurisdictions.

Acronyms and Abbreviations

A

A	Assembly (GEF)
AC	Accesion Countries
AACCLA	Latin American Association of American Chambers of Commerce
AALCC	Asian-African Legal Consultative Committee
AAPA	American Association of Port Authorities
AARINENA	Association of Agricultural Research Institutions in the Near East and North Africa
AARS	Automatic Aircraft Reporting System
AAU	Assigned Amount Unit (UNFCCC)
ABC	Brazilian Cooperation Agency
ABS	Access to genetic resources and benefit sharing (CBD)
AC (1)	Animal Committee (CITES)
AC (2)	Aårhus Convention
AC (3)	Armed Conflict
ACAP	Agreement on the Conservation of Albatrosses and Petrels
ACC	Administrative Committee on Coordination
ACCOBAMS	Agreement on the Conservation of Cetaceans of the Black Sea, the Mediterranean Sea and Contiguous Atlantic Area
ACCSWR	Administrative Committee on Coordination, Sub-Committee on Water Resources
ACDA	Arms Control and Disarmament Agency
ACI	International Cooperative Alliance
ACMED	African Centre of Meteorological Applications for Development
ACP	Africa, Caribbean, and the Pacific
ACS	Association of Caribbean States
ACSAD	Arab centre for the Studies of Arid Zones & Drylands
ACTS	African Centre for Technology Studies
ACTT	African Centre for Technology Transfer (GEF)
ADB (1)	African Development Bank (Sometimes AfDB)
ADB (2)	Asian Development Bank (Sometimes AsDB)
ADC	Andean Development Corporation
ADELA	Atlantic Community Development Group for Latin America
ADF	African Development Fund

A

ADI	Acceptable daily intake
ADPC	Asian Disaster Preparedness Center
ADR	Alternative Dispute Resolution
ADRC	Asian Disaster Reduction Center
AECI	Spanish Agency for International Cooperation
AEGDM	ASEAN Experts Group on Disaster Management
AEPS	Arctic Environmental Protection Strategy
AEWA	Agreement on the Conservation of African-Eurasian Migratory Water birds
AFD	Agence Française Développement
AfDB	African Development Bank (Sometimes ADB)
AFLEG	African Forest Law Enforcement and Governance
AFPPD	Asian Forum of Parliamentarians on Population and Development
AFSED	Arab Fund for Social and Economic Development
AFTA	ASEAN Free Trade Area
AG (1)	Assembly of Governors (IDB)
AG (2)	Australia Group
AG13	Ad hoc Group on Article 13 (UNFCCC)
AGBM	Ad hoc Group on the Berlin Mandate (UNFCCC)
AGCM	Atmospheric General Circulation Model (WMO)
AGDP	Agricultural Gross Domestic Product
AGHCL	Advisory Group on Harmonization of Classification and Labeling (OECD)
AGIRS	Agriculture Investment Research Service (IBRD)
AGO	Australian Greenhouse Office
AGR	Agriculture Department (WB)
AHTEG	Ad Hoc Technical Working Group
AIA (1)	Advance Informed Agreement (Cartagena Protocol on Biosafety)
AIA (2)	American International Association for Economic and Social Development
AIACC	Agricultural Impact Assessment of Climate Change
AIC	African Investment Association
AID	Agency for International Development (USA)
AIDIS	Pan-American Engineering Association for the Public Health and Environment
AIJ	Activities Implemented Jointly (UNFCCC)
AIMS	Atlantic, Indian Ocean, Mediteranean, and the South China Seas (SIDS grouping)
AIT	Asian Institute of Technology
ALA	EU's Assistance Programme in Asia and Latin America

ALADI	Latin American Institute for Integration and Development
ALALC	Latin American Association of Free Commerce
ALCORDES	Latin American Association of Regional Development Associations
ALESCO	Arib Centre for the Study of Arid Zones and Dry Lands (LAS)
ALIDE	Latin American Association of Development Finance Institutions
ALIDES	Central American Alliance for Sustainable Development
AMAP	Arctic Monitoring and Assessment Programme
AMCEN	African Ministerial Conference on the Environment
AMCOW	African Ministers' Council on Water
AMF	Arab Monetary Fund
ANGO	Advocacy NGO
ANPED	Northern Alliance for Sustainability
AoA	Agreement on Agriculture (WTO)
AOAD	Arab Organization for Agricultural Development
AOSIS	Alliance of Small Island States
AP (1)	Alliance for Progress
AP (2)	Associated Program
APD	Approved Project Document (IDB)
APEC	Asia-Pacific Economic Cooperation Forum
APEP	Alliance for Progress
APFED	Asia-Pacific Forum for Environment and Development
APL	Adaptable Program Lending
APO	Asian Productivity Organization
APPP	Annual Participatory Programming Process
APR	Annual Programme/Project Report (GEF)
AQUASTAT	Country Information on Water and Agriculture
AR	Agricultural Research
AR4	IPCC Fourth Assessment Report
ARC	Agricultural Research Centre
ARD	Afforestation, Reforestation, and Deforestation
ARES	African Regional Environmental Strategy
ARF	ASEAN Regional Forum
ARI	Agricultural Research Institute
ASA	Association for Social Advancement
ASCOBANS	Agreement on the Conservation of Small Cetaceans of the Baltic and North Seas
ASD	Asian Development Bank
AsDB	Asian Development Bank
ASEAN	Association of South East Asian Nations

ASIL	American Society of International Law
ASIP	Inter American Press Association
ASOEN	ASEAN Senior Officials on Environment
ATL	Technical Cooperation Loan (IDB)
ATO	African Timber Organization
AU (1)	African Union
AU (2)	Anti-bribery Undertaking
AusAid	Australian Agency for International Development
AVHRR	Advanced Very High-Resolution Radiometer
AWC	Arab Water Council
AWEC	Annual Water Experts Conference
AWF	African Water Facility (AfDB)
AWG	Advisory Working Group (CHy of WMO)
AWGB	Ad Hoc Working Group on Biodiversity (GEF/STAP)
AWGGWE	Ad Hoc Working Group on Global Warming and Energy (GEF)
AWP	Area Water Partnership (GWP)
AWTF	African Water Task Force

B

BA	Beneficiary Assessment
BACT	Best Available Control Technology
BADEA	Arab Bank for Economic Development in Africa
BAHC	Biospheric Aspects of the Hydrological Cycle
BAPA	Buenos Aires Plan of Action (UNFCCC)
BASD	Business Action for Sustainable Development
BAT	Best Available Technology
BCH	Biosafety Clearing House (CBD)
BCIS	Biodiversity Conservation Information System
BCR	Benefit-Cost Ratio
BCSD	Business Council for Sustainable Development
BE	Special Voluntary Trust Fund of the CBD
BINGO	Business and Industry NGO
BIONET	Biodiversity Action Network
BKA	Federal Chancellery (Austria)
BLICC	Business Leaders Initiative on Climate Change
BLKALET	Block A Agreement Letter (GEF)
BMA	Federal Ministry of Foreign Affairs (Austria)

BMU	German Federal Ministry of Environment
BMZ	German Federal Ministry of Development Cooperation
BNGO	Business Non-Governmental Organization
BNSC	British National Space Centre
BOD	Biological Oxygen Demand
BOG	Board of Governors
BP	Bank Procedures (WB)
BPA	Barbados Plan of Action (SIDS)
BPAO	Best Practicable Environmental Option
BPOA	Barbados Plan of Action (SIDS)
BPPE	Bureau for Programme Policy and Evaluation (GEF)
BRGM	French Geological Survey and Bureau of Mines
BSH	Basic Systems in Hydrology
BSWG	Open-ended Ad Hoc Working Group on Biosafety (CBD)
BTWC/BWC	Biological and Toxin Weapons Convention
BUWAL	Swiss Agency for Environment, Forests and Landscape
BWP (1)	Bretton Woods Project
BWP (2)	Bangladesh Water Partnership
BY	Trust Fund for the Convention (CBD)
BZ	Special Voluntary Trust Fund to Facilitate Participation by the Parties to the Convention (CBD)

C

C&I	Criteria and Indicators
CA (1)	Cooperating Agency
CA (2)	Cooperative Agreement
CA (3)	Central Administration (Italy)
CA (4)	Comprehensive Assessment of Water Management in Agriculture
CABEI	Central American Bank for Economic Integration
CAC(1)	Central American Agricultural Advisory Board
CAC(2)	Codex Alimentarius Commission
CAC&M	Central Asia, Caucasus and Moldova
CACAM	Central Asia, Caucasus, Albania and Moldova
CACM	Central American Common Market
CAD	Administrative Commission of the Executive Directorate (IDB)
CAF (1)	Andean Development Corporation
CAF (2)	Conflict Analysis Framework (WB)

CAFF	Conservation Council Conservation of Arctic Flora and Fauna
CAFTA	Central American Free Trade Agreement
CAMLR	Commission or the Conservation of Antarctic Marine Living Resources
CAMRE	Council of Arab Ministers Responsible for the Environment
CAN (1)	Climate Action Network
CAN (2)	Country Assistance Note
CANSA	Climate Action Network Southeast Asia
CANUS	Climate Action Network United States
CAP	Common Agricultural Policy
CAPAM	Commonwealth Association for Public Management
CAPNET	International Network for Capacity Building for Integrated Water Resources Management (GWP)
CAREC	Regional Development Centre for Central Asia
CARIBANK	Caribbean Development Bank
CARICOM	Caribbean Community
CARIFTA	Caribbean Free Trade Association
CARIRI	Caribbean Industrial Research Institute
CAS (1)	Country Assistance Strategy (WB)
CAS (2)	Commission for Atmospheric Sciences (WMO)
CAS (3)	Complex Adaptive Systems
CASIN	Centre for Applied Studies in International Negotiations
CAT	Convention against Torture and other Cruel, Inhuman or Degrading Treatment or Punishment
CATAC	Central American Technical Committee (GWP)
CATEP	Certified Action on Tradable Emissions Permits
CATHALAC	UNESCO Water Center for the Humid Tropics of Latin America and the Caribbean, Panama (UNESCO)
CATIE	Tropical Agriculture Training and Research Centre
CAZALAC	Water Centre for Arid and Semiarid Regions of Latin America and the Caribbean, La Serena, Chile (UNESCO)
CBD	Convention on Biological Diversity
CBDR	Common but Differentiated Responsibilities
CBH	Capacity-Building in Hydrology and Water Resources
CBI (1)	Confidential Business Information
CBI (2)	Caribbean Basin Initiative
CBNRM	Community Based Natural Resource Management
CBO	Community Based Organization
CBS	Commission for Basic Systems (WMO)
CBW	Chemical and Biological Weapons/Warfare

CCA (1)	Common Country Assessment
CCA (2)	Caribbean Conservation Association
CCA (3)	Coalition for Clean Air
CCAD	Central American Commission for Environment and Development
CCAMLR	Convention on the Conservation of Antarctic Marine Living Resources
CCAP	Climate Change Action Plan
CCC	Caribbean Conservation Corporation
CCCC	Caribbean Climate Change Centre
CCCO	Committee for Climate Changes and the Ocean
CCD	Convention to Combat Desertification
CCF	Country Cooperation Framework
CCI (1)	Commission for Climatology (WMO)
CCI (2)	Cross-Cutting Issues
CCL	Climate Change Levy
CCLM	Committee on Constitutional and Legal Matters (IU)
CCOHS	Canadian Centre for Occupational Health and Safety
CCOL	Coordinating Committee on the Ozone Layer (UNEP)
CCPR	International Covenant on Civil and Political Rights
CCPR-OP1	Optional Protocol to the CCPR
CCRH	Central American Commission for Water Resources
CCX	Chicago Climate Exchange (Market for GHG)
CD	Conference on Disarmament
CDB	Caribbean Development Bank
CDC (1)	Conservation Data Center
CDC (2)	Center for Disease Control and Prevention (USA)
CECAL	European Committee for Cooperation with Latin America
CDEF	Community Development Carbon Fund (WB)
CDERA	Caribbean Disaster Emergency Response Agency
CDF (1)	Comprehensive Development Framework (WB)
CDF (2)	Clean Development Fund (UNFCCC)
CDIAC	Carbon Dioxide Information Analysis Center
CDM	Clean Development Mechanism (UNFCCC)
CDQs	Community Development Quotas
CDRM	Comprehensive Disaster Risk Management
CE	Central Europe
CEA	Cumulative Effects Assessment
CEAC	Commission on Education and Communication (IUCN)
CEB	Chief Executives Board (UN)
CEC (1)	Commission of the European Communities

CEC (2)	Commission for Environmental Cooperation (NAFTA)
CECLA	Special Coordination Committee for Latin America (UN)
CECODHAS	European Liaison Committee for Social Housing
CECON	Special Commission for Consultation and Negotiation
CEDARE	Centre for Environment and Development in the Arab Region and Europe
CEDAW	Convention on the Elimination of all Forms of Discrimination against Women
CEE	Central and Eastern Europe
CEESP	Commission on Environmental, Economic, and Social Policy (IUCN)
CEETAC	Central and Eastern Europe Technical Advisory Committee (GWP)
CEFTA	Central European Free Trade Agreement
CEH	Centre for Ecology and Hydrology (UK)
CEHI	Caribbean Environmental Health Institute
CEITs	Countries with Economies in Transition
CEL (1)	Commission on Environmental Law (IUCN)
CEL (2)	Latin American Economic Community
CEM (1)	Country Economic Memorandum
CEM (2)	Commission on Ecosystem Management (IUCN)
CEM (3)	Committee on Economic Information and Market Intelligence (ITTA)
CENTO	Central Asian Treaty Organization
CEO (1)	Centre for Earth Observation
CEO (2)	Chief Executive Officer (GEF)
CEOP	Coordinated Enhanced Observing Period
CEOS	Committee for Earth Observation Satellites
CEP	Caribbean Environment Programme (UNEP)
CEPS	Centre for European Policy Studies
CEPES	European Centre for Higher Education (Bucharest, Romania) (UNESCO)
CEQ (1)	Commission for Environmental Quality (NAAEC)
CEQ (2)	Council of Environmental Quality (USA)
CER(1)	Certified Emissions Reduction
CER(2)	Comprehensive Evaluation Report
CERD	International Convention on the Elimination of all Forms of Racial Discrimination
CERES	Coalition for Environmentally Responsible Economies
CERN	Chinese Ecosystem Research Network

CERT	Committee on Energy Research and Technology and Working Parties (IEA)
CESCR	International Covenant on Economic, Social and Cultural Rights
CESI	Committee for Environmental and Social Impacts (IDB)
CETA	Conventional Energy Technical Assistance
CFA (1)	Concessional Finance Arrangement
CFA (2)	Committee on Finance and Administration (ITTA)
CFAW	Canadian Fund for Africa on Water
CFC	Chlorofluorocarbon
CFDT	Committee on Forest Development in the Tropics (FAO)
CFI	Committee on Forest Industry (ITTA)
CFL	Compact Fluorescent Lamp
CFR	Case Fatality Rate
CFS	Committee on World Food Security
CG (1)	Canadian Government
CG (2)	Consultative Group
CG/HCCS	Coordinating Group for the Harmonization of Chemical Classification Systems (OECD)
CG-11	Central Group-11
CGE (1)	Consultative Group of Experts (UNFCCC)
CGE (2)	Compatible General Equilibrium
CGG	Commission on Global Governance
CGIAR	Consultative Group on International Agriculture Research (IBRD)
CGLU	United Cities and Local Governance (UN)
CGMW	Commission for the Geological Map of the World
CGP	Consultative Group to Assist the Poorest (IBRD)
CGRFA	Commission on Genetic Resources for Food and Agriculture
CHARM	International Center for Water Hazard and Risk Management
CHINATAC	China Technical Advisory Committee (GWP)
CHM	Clearing-House Mechanism (SBD)
CHOGM	Commonwealth Heads of Government Meeting
CHP	Combined Heat and Power
CHR	Commission on Human Rights
CHy	Commission for Hydrology of WMO
CI (1)	Conservation International
CI (2)	Consumers International
CIAB	The Coal Industry Advisory Board (IEA)
CIAT	International Centre for Tropical Agriculture
CICERO	Centre for International Climate and Energy Research, Oslo

CICI	International Conference on the Contribution of Criteria and Indicators for Sustainable Forest Management
CIDA	Canadian International Development Agency
CIDH	Inter-American Human Rights Commission
CIDI	Inter-American Commission for Integral Development (OAS)
CIDIE	Committee of International Development Institutions on the Environment
CIDS	Inter-American Committee for Sustainable Development (OAS)
CIESIN	Consortium for International Earth Science Information Network
CIFOR	Center for International Forestry Research
CIHEAM	International Centre for Advanced Mediterranean Agronomic Studies
CILSS	Convention Establishing a Permanent Inter-States Committee for Drought Control in the Sahel
CIMMYT	International Center for the Improvement of Maize and Wheat
CIP	International Potato Center
CIS	Commonwealth of Independent States
CITES	Convention on International Trade in Endangered Species of Wild Fauna and Flora
CITO	Centre for International Training and Outreach (University of Idaho, USA)
CLAPN	Latin American Committee for National Parks
CLC	Convention on Civil Liability for Oil Pollution
CLAES	*Centro Latino Americano de Ecologia Social*
CLIVAR	Climate Variability Research Programme
CLCS	Commission on the Limits of the Continental Shelf
CMP (1)	Carbon Market Programme (UNFCCC)
CMP (2)	Conference of the Parties to the UNFCCC serving as the Meeting of the Parties to the Kyoto Protocol (UNFCCC)
CMS	Convention on the Conservation of Migratory Species of Wild Animals
CN (1)	Conference Notes
CN (2)	Committee Notes
CNA (1)	Climate Network Africa
CNA (2)	National Water Commission (Mexico)
CNE	Climate Network Europe
CNF	Canadian Nature Federation
CNMC	Committee on Non-Member Countries (IEA)
CNN	Cable Network News
CNPPA	Commission on National Parks and Protected Areas (IUCN)

CO	Country Office
COAG	Committee on Agriculture (FAO)
COD	Chemical Oxygen Demand
COFO	Committee on Forestry (FAO)
COHG	Conference of Heads of Governments of the Caribbean Community
COICA	Coordinating Body of Indigenous Organizations of the Amazon Basin
CONCAUSA	Central America/United States Joint Accord
CONF	Conference (Document Identification)
CONGO	Conference of Non-Governmental Organizations in Consultative Relationships with the United Nations
CoP (1)	Community of Practice (WB)
COP (2)	Conference of the Parties to a Convention
COP (3)	Country Operational Programme (EU)
COPUOS	Committee on the Peaceful Uses of Outer Space (UN)
CORECA	Regional Council for Agricultural Cooperation in Central America, Mexico and the Dominican Republic
CORINE	Coordination of Information on the Environment in Europe (CEC)
CO/RSAT	Operations Committee/Technical Assistance Request Summary (IDB)
CO/RSP	Operations Committee/Loan Request Summary (IDB)
COW	Committee of the Whole
CP (1)	Conference of Parties (Document identification)
CP (2)	Press Release (IDB)
CP (3)	Consulting Partners (GWP)
CPA (1)	Country Performance Assessment
CPA (2)	Loan Committee (IDB)
CPAN	Circumpolar Protected Area Network
CPACC	Caribbean Planning for Adaptation to Climate Change
CPD	Commission on Population and Development
CPF	Collaborative Partnership on Forests
CPGR/CGR	Commission for Plant Genetic Resources/Genetic Resources (FAO)
CPI (1)	Corruption Perception Index
CPI (2)	Consumer Price Index
CPIA	Country Policy and Institutional Assessment (ADB)
CPPF	Canadian Project Preparation Trust Fund (IDB)
CPR (1)	Committee of Permanent Representatives
CPR (2)	Common Pool Resources

CPR (3)	Personnel Committee (IDB)
CPR (4)	Conflict Prevention and Reconstruction Unit (WB)
CR	Country Report (GEF)
CRAMLR	Convention on the Regulation of Antarctic Mineral Resources
CRC (1)	Convention on the Rights of the Child
CRC (2)	Chemical Review Committee (Rotterdam Convention)
CRED	Centre for Research on the Epidemiology of Disasters
CRF (1)	Common Reporting Format
CRF (2)	Committee on Reforestation and Forest Management (ITTA)
CRIC	Committee for the Review of the Implementation of the Convention to Combat Desertification
CROP	Council of Regional Organizations in the Pacific
CRP	Conference Room Paper
CRRH	Regional Commission for Water Resources
CRS	Creditor Reporting System (OECD)
CSA	Cost Sharing Agreement
CSD	Commission on Sustainable Development (UN)
CSAG	Civil Society Advisory Group
CSDS	Countries in Special Development Situations (UN)
CSE	Centre for Science and Environment
CSI	Environment and Development in Coastal Regions and Small Islands (UNESCO)
CSM	Climate System Model
CSMP	Center for the Study of Marine Policy (University of Delaware)
CSO	Civil Society Organization
CSR	Civil Service Reform
CSRIO	Commonwealth Scientific and Industrial Research Organization
CST	Committee on Science and Technology (CCD)
CSW	Commission on the Status of Women
CTBT	Comprehensive Test Ban Treaty
CTE	Committee on Trade and Environment (WTO)
CTI (1)	Climate Technology Initiative (UNFCCC)
CTI (2)	Climate Technology Institute
CTIP	Climate Technology Implementation Plan (UNFCCC)
CTO	Certified Tradable Offsets (UNFCCC)
CTU	Caribbean Tourism Organization
CU (1)	Conservation Unit
CU (2)	Country Unit (IBRD)
CUS-CSD	China-US Center for Sustainable Development
CV	Curriculum Vitae

CVPU	Country Valuation Project Unit (GEF but outdated)
CWC	Chemical Weapons Convention
CWIS	Center for World Indigenous Studies
CWP	Country Water Partnership (GWP)
CY	Calendar Year

D

D&FD	Deforestation and Forest Degradation
DA	Development Assistance
DAC	Development Assistance Committee (OECD)
DALY	Disability adjusted life year
DANCED	Danish Cooperation for Environment and Development
DANIDA	Danish International Development Agency
DAW	Division for the Advancement of Women (UN)
DB	Development Business (WB)
DC	Developed Country
DCS	Data Collection System
DDC (1)	*Direction de Développement et de la Cooperation* (Switzerland)
DDC (2)	Data Distribution Center
DE	Decentralized Energy Path
DEC	Division of Environmental Conventions (UNEP)
Dec	Decision (Document Identification)
DENR	Department of Environment and Natural Resources
DEPHA	Data Exchange Platform for the Horn of Africa
DESA (1)	United Nations Department of Economic and Social Affairs
DESD	UN Decade of Education for Sustainable Development
DESIP	Demographic, Environmental, and Security Issues Project (UN)
DEWA	Division of Early Warning and Assessment (UNEP)
DFA	Department of Foreign Affairs
DFID	Department of International Development (UK)
DGCS	*Direzione Generale per la Cooperazione allo Sviluppo* (Italy)
DGD	Decision Guidance Document
DGIC	Directorate General for International Cooperation (Belgium)
DGIS	Directorate General for Development Cooperation (The Netherlands)
DHI	Danish Hydrological Institute of Water and Environment
DHS	Demographic Health Surveys

DIMP	Data and Information Management Panel (GTOS)
DINGO	Australian NGO
Distr	Distribution (documents)
DIVERSITAS	International Programme of Biodiversity Science
DL	Distance Learning
DMC	Drought Monitoring Centre
DMF	Disaster Management Facility (WB)
DMS	Double Majority Voting System
DMT	Disaster Management Team
DNA	Designated National Authority
DO	Development Objective (GEF)
DOD	Department of Defense (USA)
DOE	Department of Energy (USA)
DOEM	Designated Officials for Environmental Matters (UNEP)
DONGO	Donor-Organized NGO
DOS	Department of State (USA)
DPCSD	United Nations Department for Policy Coordination and Sustainable Development
DPI	UN Department of Public Information
DPM (1)	Disaster Preparedness and Mitigation
DPM (2)	Deputy Prime Minister
DRPC	Danube River Protection Convention
DSA	Daily Subsistence Allowance (UN)
DSB	Dispute Settlement Body (WTO)
DSD	Division for Sustainable Development (ECOSOC)
DSM	Demand-Side Management
DSS	Data Synthesis System
DTIE	Division of Technology, Industry and Economics (UNEP)
DWFNs	Distant-water Fishing Nations

E

E (1)	English
E (2)	Edited
E (3)	Executive
E/D	Environment/Development
E-7	Nine of the leading electric utilities from the G-7 countries
EA	Environmental Assessment

EA-5	East Asia 5 (Indonesia, Korea, Malaysia, Philippines, Thailand)
EA	Executing Agency (GEF)
EAC	European Advisory Committee
EACC	East African Cooperation Community
EAGs	Environmental Assessment Guidelines
EA-RWP	Eastern Africa Regional Water Partnership
EAS-RCU	East Asian Seas Regional Coordinating Unit (UNEP)
EATRR	European Agreement on the Transfer of Responsibility for Refugees
EB	Executive Board
EBRD	European Bank for Reconstruction and Development
EC (1)	European Community
EC (2)	Commission of the European Communities
ECA (1)	United Nations Economic Commission for Africa
ECA (2)	Export Credit Agency
ECBA	Economic Cost Benefit Analysis
ECE	United Nations Economic Commission for Europe
ECEH	European Centre for Environment and Health
ECG	Ecosystem Conservation Group (UNEP, FAO, UNESCO, IUCN, UNDP, WWF)
ECHO	Humanitarian Office of the European Community
ECI	Earth Charter Initiative
ECLAC	United Nations Economic Commission for Latin American and the Caribbean
ECN (1)	Energy Research Centre of the Netherlands
ECN (2)	Environmental Change Network
ECO (1)	Earth Council Ombudsman (Earth Council)
ECO (2)	Pan American Center for Human Ecology and Health (PAHO)
ECO (3)	Ecological Citizens Organization
ECO (4)	Environmental Community Organization
ECOLEX	Environmental Law Information System (UNEP, IUCN, FAO)
ECON	Ministry of Economy and Finance (Spain)
ECOSOC	United Nations Economic and Social Council
ECOWAS	Economic Commission of West African States
ECPF	Expert Consultation on the Role of Planted Forests (FAO)
ED	Executive Director
EDF	European Development Fund
EDI	Economic Development Institute (Now WBI)
EDUC	Ministry of Education and Science (Spain)
EE&C	Environmental Education and Communication

EEA (1)	European Environment Agency
EEA (2)	European Economic Area
EEAD	Environmental Effects Assessment Panel
EEAS	Energy Efficiency Accreditation Scheme
EEB	European Environmental Bureau
EEOS	European Earth Observing System
EEWP	Energy Efficiency Working Party (IEA)
EEZ	Exclusive Economic Zone
EfE	Environment for Europe
EFI	European Forest Institute
EFTA	European Free Trade Association
EGM	Expert Group Meeting
EGTT	Expert Group on Technology Transfer
EHC	Environmental Health Criteria (IPCS)
EIA	Environmental Impact Assessment
EIB	European Investment Bank
EIG	Environmental Integrity Group
EIR	Extractive Industries Review
EIS	Environmental Impact Statement
EIT	Economies in Transition Countries
ELANEM	Euro-Latin American Network for Environmental Assessment and Monitoring
ELC	Environmental Law Centre (IUCN)
ELCI	Environmental Liaison Centre International
ELS	Ecolabelling Schemes
EMA	Emissions Marketing Association
EMAN	Ecological Monitoring and Assessment Network
EMAP	Environmental Monitoring and Assessment Programme
EMAS	Eco-management and Audit Scheme
EMB	Environmental Management Bureau
EMDG	Emissions Market Development Group
EMEP	Co-operative Programme for Monitoring and Evaluation of the Long-Range Transmission of Air Pollutants in Europe
EMF	Environmental Managemant Foundation (Holland)
EMG	Environmental Management Group
EMP	Euro-Mediterranean Partnership
EMS	Environmental Management System
EMU	European Monetary Union
ENB	Earth Negotiations Bulletin

ENCORE	Eastern Caribbean Environment and Coastal Resources Management Project
ENDA-TM	Environment and Development Action in the Third World
ENGs	Electronic Navigational Charts
ENGO	Environmental non-governmental organization
ENN	Environmental Network News
ENRG	Environment and Natural Resources Group (UNDP)
ENRICH	European Network for Research in Global Change
ENSO	El Niño Southern Oscillation
ENTRI	Environmental Treaties and Resource Indicators
ENV (1)	Environment Department (WB)
ENV (2)	Ministry of Environment (Spain)
ENVGC	Global Environment Coordination Division (WB)
EO	Earth Observation
EOS	Earth Observatory Satellite
EOSG	Executive Office of the Secretary General (UN)
EOU	Evaluation and Oversight Unit (UNEP)
EP	Environmental Programme
EPA	United States Environmental Protection Agency
EPC	European Patent Convention
EPD	Environmental Product Declaration
EPE	European Partners for the Environment
EPOCH	European Programme on Climatology and Natural Hazards
EQS	Environmental Quality Standards
ERA	European Research Area
ERO	Expert Advisory Panel on Emergency Relief Operations (WHO)
EROS	US Geological Survey Earth Resources Observation Systems
ERSDAC	Earth Remote Sensing Data Analysis Center
ERT	Expert Review Team
ERU	Emissions Reduction Unit (UNFCCC)
ES	Environmental Statement
ESA (1)	European Space Agency
ESA (2)	Ecological Society of America
ESA (3)	Environmental and Social Assessment
ESA (4)	Environmentally Sensitive Area
ESA (5)	Economic Stakeholder Analysis
ESAF	Enhanced Structural Adjustment Facility
ESAP	Environmental and Social Assessment Procedure
ESCAP	United Nations Economic and Social Commission for Asia and the Pacific

ESCWA	United Nations Economic and Social Commission for Western Asia
ESD (1)	Environmentally and Socially Sustainable Development Network (WB)
ESD (2)	Education for Sustainable Development
ESDG	Environmental Sustainable Development Group
ESIA	Environmental and Social Impact Assessment
ESMP	Environmental and Social Management Plan (ADB)
ESP	Environmental and Social Policy Papers
ESPAR	Agricultural Research Group of the World Bank
ESTs	Environmentally Sustainable Technologies
ESW	Economic and Sector Work (WB)
ET	Emissions Trading (UNFCCC)
ETPA	Education, Training and Public Awareness (UNFCCC)
ETS	Emissions Trading Scheme
EU (1)	European Union
EU (2)	European Community and its member states
EUFORGEN	European Forest Genetic Resources Programme
EUROBATS	Agreement on the Conservation of Bats in Europe
EUROPARL	European Parliament
EUROSTAT	Statistical Office of the European Communities
EUWI	European Union Water Initiative
EVI	Environment Vulnerability Index
EWI	Ecosystem Wellbeing Index
EXC (1)	Executive Council (GEF)
EXC (2)	Office of the President (WB)
EXCOM	Executive Committee (Multilateral Fund for the Montreal Protocol)
EXCOP	Extraordinary Meeting of the Conference of the Parties
EXIM	Export-Import Bank of the United States

F

FAO	Food and Agriculture Organization (UN)
FBD	Forest Biodiversity
FCB	Fuel Cell Buses
FCCC	Framework Convention on Climate Change (also UNFCCC)
FCM	Forest Concession Management

F

FCZ	Fishery Conservation Zone
FDA	Food and Drug Administration (USA)
FDI	Foreign Direct Investment
FEPS	Final Executive Project Summary (GEF)
FEWS	Famine Early Warning Systems Network
FFA (1)	Forum Fisheries Agency
FFA (2)	Framework for Action (GWP)
FFC	Forum Fisheries Committee
FfD	Financing for Development (also FFD)
FFEM	French Fund for the Global Environment
FHI	Family Health International
FI	Financial Initiative (UNEP)
FIDA	Inter-American Forum on Environmental Law
FIELD	Foundation of International Law Development
FINNIDA	Finnish International Development Agency
FIS	International Seed Trade Federation
FLACSO	Latin American Social Science Faculty
FLEG	Forest Law Enforcement and Governance
FLR	Forest Landscape Restoration
FMU	Forest Management Unit
FO	Foreign Office
FOC	Friends of the Chair
FOEI	Friends of the Earth International
FORNESSA	Forestry Research Network for Sub-Saharan Africa
FOSA	Forestry Outlook Study for Africa
FP	Family Planning
FPG	Finance Partners Group (GWP)
FRA	Forest Resource Assessment
FRIEND	Flow Regimes from International Experimental and Network Data (UNESCO)
FSC	Forest Stewardship Council
FSO	Fund for Special Operations (IDB)
FSP	*Fonds de solidarité prioritaire* (France)
FT	Fuzzy thinking
FTA	Free Trade Area of the Americas
FTAA	Free Trade Area
FTP	Forest, Trees, and People Programme (FAO)
FWCW	Fourth World Conference on Women
FY	Fiscal Year

G

G

G	General Distribution (Document Identification)
G-7	Group of seven (United States, Japan, Germany, France, Britain, Italy, and Canada)
G-8	Group of eight (G-7 and Russia)
G-15	Group of fifteen (Summit level group of developing countries)
G-77/China	Group of 77 developing countries plus Chine (Now includes nearly 150 governments)
GA	UN General Assembly
GAIM	Global Analysis and Modeling
GATS	General Agreement on Trade in Services
GATT	General Agreement on Tariffs and Trade
GAW	Global Atmospheric Watch
GBF	Global Biodiversity Forum
GBIF	Global Biodiversity Information Facility
GC	Governing Council
GC/GMEF	Governing Council/Global Ministrial Environment Forum
GCA	Global Coalition for Africa
GCC (1)	Global Climate Change
GCC (2)	Global Climate Coalition
GCC(3)	Gulf Cooperation Council
GCI	Green Cross International
GCM	General Circulation Model
GCOS	Global Climate Observing System
GCP	Government Cooperative Programme (FAO)
GCRIO	US Global Change Research Information Office
GCRMN	Global Coral Reef Monitoring Network
GCTE	Global Change and Terrestrial Ecosystems
GD	Global Deal
GDI	Gender-related Development Index
GDLN	Global Development Learning Network (World Bank)
GDP	Gross Domestic Product
GEA	Global Environmental Action
GECHS	Global Environmental Change and Human Security
GEF	Global Environment Facility
GEFIA	Global Environment Facility Interim Approval
GEF-IN	Global Environment Facility International Notes
GEF-IW	Global Environment Facility International Waters

G

GEFOP	Global Environment Facility Operations Committee
GEFPID	Project Information Document (GEFWB)
GEFRC	Global Environment Facility Regional Coordinator
GEFSEC	Global Environment Facility Secretariat
GEFTF	Global Environment Facility Trust Fund
GEMI	Global Environmental Management Initiative
GEMS	Global Environment Monitoring System (UNEP)
GEO	Global Environment Outlook Report (UNEP)
GESAMP	Group of Experts on the Scientific Aspects of Marine Pollution
GEST	Global Evaluation of Sediment Transport
GET	Global Environment Trust Fund
GETF	Global Environment Technology Foundation
GEWEX	Global Energy and Water Cycle Experiment
GFAR	Global Forum on Agricultural Research
GFMC	Global Fire Monitoring System
GFRA	Global Forest Resources Assessment
GFSE	Global Forum on Sustainable Energy
Gg	Gigagram
GHG	Greenhouse Gas
GHP	GEWEX Hydrometeorology Panel
GIBIP	Green Industry Biotechnology Platform
GIDP	Gender in Development Programme (UNDP)
GIFAR	Global Forum on Agricultural Research
GIPME	Global Investigation on Pollution in the Marine Environment
GIS	Geographic Information System
GISD	Geographic Information for Sustainable Development
GISP	Global Invasive Species Programme
GIWA	Global International Waters Assessment (GEF)
GLOBE	Global Legislators Organization for a Balanced Environment
GM (1)	Genetically Modified
GM (2)	Global Mechanism
GMA	Global Assessment on the Marine Environment
GMEF	Global Ministerial Environment Forum
GMES	Global Monitoring for Environment and Security
GMO	Genetically Modified Organism
GMSR	Greater Mekong (River) Sub Region
GNP	Gross National Product
GO	Global Environmental Objective (GEF)
GOFC	Global Observation of Forest Cover
GONGO	Government-organized NGO

G

GOOS	Global Ocean Observing System
GOS	Global Observing System (WMO)
GOs	Governmental Organizations
GOSIC	Global Observation Systems Information Center
"GOUTHE"	Global Observatory of Units for Teaching, Training and Ethics (UNESCO)
GP (1)	Government of Portugal
GP (2)	Good practice
GPA	Global Programme of Action
GPCC	Global Precipitation and Climatology Centre
gpgNet	Global Public Goods Network
GPG (1)	Global Public Goods
GPG (2)	Good Practice Guide
GPS	Global Positioning System
GRAIN	Genetic Resources Action International
GRDC	Global Runoff Date Center
GRI	Global Reporting Initiative
GRID	Global Resource Information Database (UN)
GRINGO	Government-run NGO
GRO	Grassroots Organization
GROMS	Global Register of Migratory Species
GroWI	Global Review of Wetland Resources and Priorities for Wetland Inventory
GRSO	Grassroots Support Organization
GRULAC	Latin American and Caribbean Group
GS	Gold Standard
GSCO	Global Social Change Organization
GSFC	Goddard Space Flight Center
GSTP	Global System of Trade Preferences (UNCTAD)
GTI	Global Taxonomy Initiative
GIN-H	Global Terrestrial Network on Hydrology
GSPC	Global Strategy for Plant Conservation (CBD)
GTOS	Global Terrestrial Observing System
GTRZ	German Association for Technical Cooperation
GTSC	GTOS Steering Committee
GURTs	Genetic Use Restriction Technologies
GW	Global Warming
GWA	Gender Water Alliance
GWP (1)	Global Warming Potential (UNFCCC)
GWP (2)	Global Water Partnership

GWP-TAC	Global Water Partnership -Technical Advisory Committee 2000
GWPO	Global Water Partnership Organization
GWP-CATAC	Central American Technical Advisory Committee to the GWP
GWP-SAMTAC	South American Technical Advisory Committee to the GWP
GWS	George Wright Society
GWSATC	Global Water Science, Assesment and Training Center
GYF	Global Youth Forum (UNEP)

H

HABITAT	United Nations Centre for Human Settlement
HCFC	Halogenated CFCs
HCFCs	Hydrochlorofluorcarbons
HD	Human Development Network
HDI	Human Development Index
HDP	Human Dimensions of Global Environmental Change Programme
HELCOM	Baltic Marine Environment Protection Commission (Helsinki Commission)
HELP	Hydrology for the Environment, Life and Policy
HEM	Harmonization of Environmental Measurement Unit
HFCs	Hydro fluorocarbons
HG	Head of Government
HIA	Health Impact Assessment
HI/HYVs	High-Input and High Yield Hybrid Crop Varieties
HIPC	Heavily Indebted Poor Countries
HIV/AIDS	Human Immune-Deficiency Virus
HLCOMO	High-level Committee of Ministers and Officials (UNEP)
HNP	Health, Nutrition and Population
HNS	International Convention on Liability and Compensation for Damage in Connection with the Carriage of Hazardous and Noxious Substances (IMO)
HNS	Hazard and Noxious Substances (IMO)
HPI	Human Poverty Index
HQ	Headquarters
HRC	Human Rights Commission
HRD	Human Resources Development
HRI	Hydraulic Research Institute (Egypt)
HS	Head of State

HST	Hubble Space Telescope
HTS	Harmonized Tariff Schedule
HWI	Human Wellbeing Index
HWP	Harvested Wood Products (UNFCCC)
HWRP	Hydrology Water Resources Programme

I

IA (1)	Implementing Agency (GEF)
IA (2)	Implementing Arrangements (IEA)
IA (3)	Institutional Analysis (WB)
IAB	Industry Advisory Board (IEA)
IABIN	Inter-American Biodiversity Information Network
IAC	Informal Advisory Committee
IACD	Inter-American Agency for Cooperation and Development (OAS)
IACNDR	Inter-American Committee for Natural Disaster Reduction
IACSD	UN Interagency Committee on Sustainable Development
IADB (1)	Inter-American Development Bank (IDB)
IADB (2)	Inter-American Defense Board
IADRIP	Inter-American Declaration on the Rights of Indigenous Peoples
IADWM	Inter-American Dialogue on Water Management
IAEA	International Atomic Energy Agency (UN)
IAEC	Inter-American Economic Council
IAF	Inter-American Foundation
IAH	International Association of Hydrologists
IAHR	International Association of Hydraulic Engineering and Research
IAHS	International Association of Hydrological Sciences
IAI	Inter-American Institute for Global Change Research
IAIA	International Association for Impact Assessment
IAP-WASAD	International Action Programme on Water and Sustainable Agricultural Development (FAO)
IARC	International Agriculture Research Center
IASI	Inter-Americas Sea Initiative
IASWS	International Association for Sediment and water Science
IATF	Interagency Task Force on Sustainable Development (OAS)
IATTC	Inter-American Tropical Tuna Commission
IBA	International Bird Area
IBAMA	Institute of Environment and Renewable Resources (Brazil)

IBE	International Bureau of Education (Geneva, Switzerland) (UNESCO)
IBPGR	International Board for Plant Genetic Resources (CGIAR)
IBRD	International Bank for Reconstruction and Development
IBTA	Institution-building/Technical Assistance
IC (1)	Implementation Committee (GEF/STAP)
IC (2)	Initial Communication
IC (3)	Inter-Sessional Committee
ICAD	Integrated Conservation and Development Project (GEF)
ICALPE	International Centre for Alpine Environments
ICAO	International Civil Aviation Organization
ICARDA	International Center for Agricultural Research in the Dry Areas
ICARM	Integrated Coastal and River Basin Management
ICASE	International Council of Associations for Science Education
ICC (1)	International Chamber of Commerce
ICC (2)	International Criminal Court
ICCAT	International Convention on the Conservation of the Atlantic Tuna
ICCBD	Intergovernmental Committee on the Convention on Biological Diversity
ICCD (1)	Intergovernmental Committee on the Convention to Combat Desertification
ICCD (2)	Intergovernmental Climate Change Directorate
ICCE	IAHS International Commission on Continental Erosion
ICCP	Intergovernmental Committee on the Cartagena Protocol on Biosafety (CBD)
ICCPR	International Covenant on Civil and Political Rights
ICCROM	International Centre for the Study of the Preservation and Restoration of Cultural Property
ICWHRM	International Centre for Water Hazard and Risk Management (Japan/UNESCO)
ICDA	International Coalition of Development Action
ICDI	Independent Commission on International Development Issues (The Brandt Commission)
ICDP	Integrated Conservation Development Project (GEF)
ICES	International Council for the Exploration of the Sea
ICESCR	International Covenant on Economic, Social, and Cultural Rights
ICFD	International Conference on Financing for Development
ICID	International Commission on Irrigation and Drainage
ICIMOD	International Centre for Integrated Mountain Development

ICJ	International Court of Justice
ICLARM	International Centre for Living Aquatic Resources Management
ICLEI	International Council for Local Environmental Initiatives
ICM(1)	Integrated Coastal Management
ICM(2)	International Conflict Management
I-CMAP	First Mesoamerican Congress on Protected Areas
ICM	Information and Communication Management
ICO	*Instituto de Crédito Oficial*
ICOH	International Commission on Occupational Health
ICOLD	International Commission on Large Dams
ICOMOS	International Council of Monuments and Sites
ICP	Informal Consultative Process on Oceans and the Law of the Sea
ICPD	International Conference on Population and Development
ICPRD	International Commission for the Protection of the Danube River
ICPs	International Cooperative Programmes
ICRAF	International Centre for Research in Agroforestry
ICRAN	International Coral Reef Action Network
ICRC (1)	Interim Chemical Review Committee (PIC)
ICRC (2)	International Committee of the Red Cross
ICRG	International Consulting Resources Group
ICRI	International Coral Reef Initiative
ICRISAT	International Crops Research Institute for the Semi-Arid Tropics
ICSID	International Centre for Settlement of Investment Disputes (WB)
ICSU	International Council of Scientific Unions
ICTs	Information Communication Technologies
ICTP	International Centre for Theoretical Physics (Trieste, Italy) (UNESCO)
ICWE	International Conference on Water and the Environment
ICZM	International Coastal Zone Management
IDA	International Development Association (WB)
IDAC	International Disaster Advisory Committee
IDB (1)	Inter-American Development Bank
IDB (2)	Inter-American Defense Board
IDB (3)	Islamic Development Bank
IDF	Institutional Development Facility
IDGEC	Institutional Dimensions of Global Environmental Change
IDGs	International Development Goals
IDNDR	International Decade for Natural Disaster Reduction
IDP (1)	Integrated Development Plan
IDP (2)	Internal Discussion Paper (WB)

IDR	In-depth Review (UNFCCC)
IDRC	International Development Research Centre
IDU	Internal Document Unit (WB)
IEA	International Energy Agency
IEEP	Institute for European Environment Policy
IEG	International Environmental Governance
IEGS	Institute for Global Environmental Strategies
IEHS	Institute for Environment and Human Security (UNU)
IEO	Industry and Environment Office (UNEP)
IEPS	Initial Executive Project Summary (GEF)
IESALC	International Institute for Higher Education in Latin America and the Caribbean (Caracas, Venezuela) (UNESCO)
IET	International Emissions Trading (UNFCCC)
IETA	International Emissions Trading Association
IETC	International Environmental Technology Centre
IFAD	International Fund for Agricultural Development
IFAW	International Fund for Animal Welfare
IFC	International Finance Corporation (WB)
IFCS	International Forum on Chemical Safety
IFEJ	International Federation of Environmental Journalists
IFF (1)	Intergovernmental Forum on Forests
IFF (2)	International Finance Facility
IFI	International Financing Institution
IFM	Integrated Flood Management
IFNet	International Flood Network and Flood Alert System
IFPRI	International Food Policy Research Institute
IFRC	International Federation of Red Cross and Red Crescent Societies
IG	Inter-governmental
IGBP	International Geosphere-Biosphere Programme
IGC (1)	International Green Cross
IGC (2)	Intergovernmental Consultation
IGES	Institute for Global Environmental Strategies (Japan Environment Agency)
IGFA	International Group of Funding Agencies for Global Change
IGM	Open-ended Intergovernmental Group of Ministers or their Representatives
IGO	Inter-Governmental Organization
IGOS	Integrated Global Observing Strategy
IGR	Intergovernmental Review

IGRAC	International Groundwater Resources Assessment Centre, Utrecht, The Netherlands (UNESCO; WHO)
IHD	International Hydrological Decade
IHDP	International Human Dimensions Programme on Global Change
IHE-Delft	International Institute for Infrastructural, Hydraulic and Environmental Engineering (1957-2003, Holland)
IHO	International Hydrographic Organization
IHP	International Hydrological Programme (UNESCO)
IHP-V	Fifth Phase of IHP (1996-2001)
IHP-VI	Sixth Phase of IHP (2002-2007)
IHP-VII	Seventh Phase of IHP (2008-2013)
IIASA (1)	International Institute for Applied Systems Analysis
IIASA (2)	Institute of Islamic and Arabic Sciences in America
IICBA	International Institute for Capacity-Building in Africa (Addis Ababa, Ethiopia) (UNESCO)
IICA	Inter-American Institute for Cooperation on Agriculture
IICD	International Institute for Communication and Development
IIDEA	International Institute for Democracy and Electoral Assistance
IIE	Institute for International Economics
IIED	International Institute for Environment and Development
IIEP	International Institute for Educational Planning (Buenos Aires, Argentina) (UNESCO)
IIFB	International Indigenous Forum on Biodiversity
IIMI	International Irrigation Management Institute
IIP	International Implementation Priorities
IISD	International Institute for Sustainable Development (Canada)
IITA	International Institute of Tropical Agriculture
IITE	Institute for Information Technologies in Education (Moscow, Russian Federation) (UNESCO)
IJC	International Joint Commission
ILAC	Latin American and Caribian Initiative for Sustainable Societies
ILEC	International Lake Environment Committee
ILO	International Labor Organization (UN)
ILRAD	International Laboratory for Research on Animal Disease
ILRI	International Livestock Research Institute
ILSI	International Life Sciences Institute
ILTER	International Long-Term Ecological Research
IMCAM	Integrated Marine and Coastal Area Management
IMCO	Intergovernmental Maritime Consultative Organization
IMF	International Monetary Fund

IMO	International Maritime Organization
IMP	Information Management Plan
INBAR	International Network for Bamboo and Ratan
INBO	International Network of Basin Organizations
INC	International Negotiation Committee
INC/FCCC	Intergovernmental Negotiation Committee for a Framework Convention on Climate Change
INCCC/INC/FCCC	Intergovernmental Negotiating Committee for a Framework Convention on Climate Change
INCD	International Negotiation Committee on Desertification
INDOEX	Indian Ocean Experiment
INENCO	Centre for International Environmental Cooperation
Inf	Information (Documents Identification)
INFOCAP	Information Exchange Network on Capacity Building for the Sound Management of Chemicals
INFOTERRA	International Environmental Information System
INGO (1)	Individual-based NGO
INGO (2)	International NGO
INPE	National Institute for Space Research (Brazil)
INSERM	French National Institute for Health and Medical Research
INSTRAW	International Research and Training Institute for the Advancement of Women (UN)
INTAL	Institute for the Integration of Latin America and the Caribbean
INTECOL	International Association of Ecology
INV00	Inventory Submission for the Year 2000
IO (1)	International Organization
IO (2)	Input-Output
IOC	Intergovernmental Oceanographic Commission (UNESCO)
IOCC	Inter-Organization Coordinating Committee
IOCU	International Organization of Consumers Unions
IOI	International Oceans Institute
IOSEA	Indian Ocean-South East Asia
IP	Implementation Progress (GEF)
IPCC	Intergovernmental Panel on Climate Change (UNEP/WHO)
IPDP	Indigenous Peoples Development Plan (WB)
IPE	Independent Panel of Experts
IPED	International Panel of Experts on Desertification
IPF (1)	Intergovernmental Panel on Forests
IPF (2)	Indicative Planning Figure (UNDP)
IPGR	International Plant Genetic Resources Institute

IPGRI	International Plant Genetic Resources Institute
IPM	Integrated Pest Management
IPO	Indigenous Peoples Organization
IPPC	International Plant Protection Convention
IPPF	International Planned Parenthood Federation
IPR	Intellectual Property Right
IPU	Inter-Parliamentary Union
IQC	Indefinite Quantity Contract (USAID)
IRBM	Integrated River Basin Management
IRC	International Research Centre for Water Supply and Sanitation
IRD	Research Institute for Development
IRI	International Resource Institute
IRPTC	International Register of Potentially Toxic Chemicals
IRR	Internal Rate of Return
IRRI	International Rice Research Institute
IRIS	Institute for Research and Innovation in Sustainability (Canada)
IRTCES	International Research and Training Centre on Erosion and Sedimentation, Beijing, China (UNESCO)
IRTCUD	International Research and Training Centre on Urban Drainage, Kuala Lumpur, Malaysia (UNESCO)
IRWR	Internal renewable water resource
ISA	International Seabed Authority
ISARM	International Shared Aquifer Resource Management
ISBN	International Standard Book Numbers
ISDF	International Sustainable Development Foundation
ISDR	International Strategy for Disaster Reduction (UN)
ISESCO	Islamic Educational , Scientific and Cultural Organization
ISI	International Sedimentation Initiative
ISIC	International Standard Industrial Classification
ISNAR	International Service for National Agricultural Research
ISO	International Organization for Standardization
ISOC	Inter-Sessional Meeting on the Operations of the Convention
ISP (1)	Inter-American Strategy for Public Participation in Decision-making for Sustainable Development
ISP (2)	Integral Sustainable Production Units
ISRIC	International Soil and Reference Information Center
ISS	Institute of Social Studies (Holland)
ISSC	International Social Science Council
ISSN	International Standard Series Numbers
IT	Information Technology

ITC (1)	International Institute for Geo-Information Science and Earth Observation (Holland)
ITC (2)	International Trade Centre
ITDG	Intermediate Technology Development Group
ITE	Institute for Terrestrial Ecology
ITFF	Interagency Task Force on Forests
ITLOS	International Tribunal for the Law of the Sea
ITOPF	International Tanker Owners pollution Federation
ITPGR(FA)	International Treaty on Plant Genetic Resources for Food and Agriculture (Formerly the International Undertaking)
ITQs	Individual Transferable Quotas
ITR	Independent Technical Review (GEF)
ITTA	International Tropical Timber Agreement
ITTC	International Tropical Timber Council
ITTO	International Tropical Timber Organization
IU	International Undertaking (CGRFA)
IUCN	World Conservation Union (International Union for the Conservation of Nature and Natural Resources)
IUFRO	International Union of Forestry Research Organizations
IUPGR	International Undertaking on Plant Genetic Resources
IUU	Illegal, unregulated, unreported
IU-WG	International Undertaking Working Group
IVA	Industrial Value Added
IW	International Waters Focal Area of the GEF
IWA	International Water Association
IWC	International Whaling Commission
IWGMP	Intergovernmental Working Group on Marine Pollution
IW-Learn	International Water Learning Exchange and Resource Network (UNDP)
IWM	Institute of Water Modelling (Bangladesh)
IWMI	International Water Management Institute
IWP	International Waters Protection
IWR	Institute for Water Resources (USACE)
IWRA	International Water Resource Association
IWRB	International Waterfowl Research Bureau
IWRM	Integrated Water Resource Management
IWRN	Inter-American Water Resources Network
IWSD	Institute of Water and Sanitation Development
IYM	International Year of Mountains
IYFW	International Year of Freshwater (2003)

J

JAC	Joint Advisory Committee
JANA	Japan Water Resource Association
JAROS	Japan Resources Observation System Organization
JAWA	Japan Water Resource Association
JBIC	Japan Bank for International Cooperation
JCEDAR	Joint Committee on Environment and Development in the Arab Region
JI	Joint Implementation (UNFCCC)
JICA	Japan International Cooperation Agency
JIN	Joint Implementation Network
JIQ	Joint Implementation Quarterly
JIU	Joint Inspection Unit
JMP	Joint Monitoring Programme
JMPR	Joint Meeting on Pesticide Residues (FAO/WHO)
JPAC	Joint Public Advisory Committee (CEC)
JPO	Junior Professional Officer
JPOI	Johannesburg Plan of Implementation
JRC	Joint Research Council
JSTC	Joint Scientific and Technical Committee (GCOS)
JUSSCANNZ	Japan, US, Switzerland, Canada, Australia, Norway and New Zealand (The non-EU industrialized countries)
JWG	Joint Working Group on Compliance (UNFCCC)

K

KFW	*Kreditanstalt für Wiederaufbau* (Germany)
KIT	Royal Tropical Institute (Netherlands)
KMOE	Korean Ministry of Environment
KMS	Knowledge Management System
KNMI	Royal Netherlands Meteorological Institute
KOWALCO	Korean Water Resources Corporation

L

L	Limited Distribution (UN Document Identification)
LA	Local Administration (Italy)
LA21s	Local Agendas 21
LAC	Latin America and the Caribbean
LAIA	Latin American Integration Association
LANBO	Latin American Network of Basin Organizations
LAS	League of Arab States
LBA (1)	Land-Based Activities
LBA (2)	Large Scale Biosphere-Atmosphere Experiment in Amazonia
LBI	Legally Binding Instrument
LBS	Land-Based Sources of Pollution
LC (1)	Latin America and the Caribbean (documents)
LC (2)	Life Cycle
LCA	Life Cycle Assessments
LCIA	Life Cycle Impact Assessments
LCSES	Latin American and Caribbean Environmentally and Socially Sustainable Development Unit (IBRD)
LDC	Least Developed Country
LEAD	Leadership for Environment and Development
LEG	Legal Department (GEF-WB)
LEGEN	Environment Negotiations Unit of the Legal Department (WB)
LFCCs	Low Forest Cover Countries (UNFF)
LG	Liaison Group
LIL	Learning and Innovation Loan
LISA	Low-Impact Sustainable Agriculture
LLDC	Least Developed Countries
LME	Large Marine Ecosystems
LMMC	Like-Minded Megadiverse Countries
LMO	Living Modified Organism (CBD)
LOA	Letter of Agreement (GEF)
LogFrame	Logical Framework
LOICZ	Land-Ocean Interaction in the Coastal Zone
LOS	Law of the Sea Convention
LP	Loan Proposal (IDB)
LRT	Long-Range Transportation
LRTAP	Convention on Long-Range Transboundary Air Pollution

LTAs	Long-Term Agreements (IEA)
LTER	Long-Term Ecological Research
LUCC	Land-Use and Land-Cover Change (IGBP/HDP)
LUCF	Land Use Change and Forestry (UNFCCC)
LULUCF	Land Use, Land Use Change, and Forestry (UNFCCC)

M

M&E	Monitoring and Evaluation
MA	Millennium Ecosystem Assessment
MAB	Man and the Biosphere Programme (UNESCO)
MAC (1)	Maximum Allowable Concentration
MAC (2)	Marginal Abatement Costs
MAC(3)	Marine Aquarium Council
MAD	Marginal Avoided Damage Costs
MAE	*Ministére des Affaires Étrangéres* (France)
MAI	Multilateral Agreement on Investment
MAP (1)	Mediterranean Action Plan
MAP (2)	Millennium African Programme
MAR	Monitoring, Assessment and Reporting
MARPOL	International Convention for the Prevention of Pollution from Ships
MAT	Mutually Agreed Terms
MB	Methyl Bromide
MC	*Mediocrecito Centrale* (Italy)
MCED	Ministerial Conference on Environment and Development
MCPFE	Ministerial Conference on the Protection of Forests in Europe
MCS	Monitoring, Control and Surveillance
MCSD	Mediterranean Commission for Sustainable Development
MDB	Multilateral Development Bank
MDF	Mediterranean Development Forum
MDGs	Millennium Development Goals
MDI	Multilateral Development Institution
MDIs	Material-dose Inhalers
MEA	Multilateral Environmental Agreement
MEDA	Mediterranean Economic Development Assistance
MEDIES	Mediterranean Education Initiative for Environment and Sustainability

MEDPOL	Programme for the Assessment and Control of Pollution in the Mediterranean Region
MEDTAC	Mediterranean Technical Committee (GWP)
MEP	Member European Parliament
MER	Market Exchange Rates
MERCOSUR	Southern Common Market (Argentina, Brazil, Chile, Paraguay, Uruguay)
MES	Markets for Environmental Services
MSSD	Mediterranean Strategy for Sustainable Development
METI	Ministry of Economy, Trade and Industry of japan
MF (1)	Multilateral Fund
MF (2)	Ministry of Finance (Belgium)
MFA	Ministry of Foreign Affairs
MFMP	Multilateral Fund for the Montreal Protocol
MFN	Most Favored Nation Status
Mg	Megagram
MHLC	Multilateral High-Level Conference
MICS	Multiple Indicator Cluster surveys
MIF	Multinational Investment Fund (IDB)
MIGA	Multilateral Investment Guarantee Agency (WB)
MIS (1)	Management Information System
MIS (2)	Methods, Inventory, and Science Programme (UNFCCC)
MLF	Multilateral Fund
MMSD	Mining, minerals and sustainable development
MMTCDE	Million Metric Tons of Carbon Dioxide Equivalents
MNC	Multinational Corporation
MoD	Ministry of Defense
MODIS	Moderate resolution imaging spectroradiometer
MOFA	Ministry of Foreign Affairs
MOI	Memorandum of Intent
MOJ	Ministry of Justice
MOP (1)	Meeting of the Parties (for a Protocol)
MOP (2)	Memorandum and Recommendation of the President (WB)
MOST	Management of Social Transformations
MOU	Memorandum of Understanding
MP (1)	Montreal Protocol
MP (2)	Member of Parliament
MPA	Marine Protected Area
MPD	Maximum Permissible Doses
MRA	Mutual Recognition Agreement

M

MRC	Mekong River Commission
MRI	Mitsubishi Research Group
MS (1)	Multilateral System (CGRFA)
MS (2)	Multistakeholder
MSC	Marine Stewardship Council
MSD	Multi-Stakeholder Dialogue
MSF	Doctors without Borders
MSP	Medium Sized Project
MSTCDE	Million Short Tons of Carbon Dioxide Equivalents
MTA	Material Transfer Agreement
MTCR	Missile Technology Control Regime
MTNs	Multilateral Trade Negotiations
MtPA	Mountain Protected Area
MTP	Medium Term Plan
MTPW	Medium-Term Programme of Work
MTS	Multilateral Trading System
MUSE	Multilateral System for Exchange
MVP	Minimum Viable Population
MWC	International Convention on the Protection of the Rights of all Migrant Workers and Members of their Families
MYPOW	Multi-year Programme of Work

N

N2O	Nitrous Oxide
NA	North America
NAAEC	North American Agreement on Environmental Cooperation (NAFTA)
NABIN	North American Biodiversity Information System
NAC	New Agenda Coalition
NACEC	North American Commission for Environmental Cooperation
NAFEC	North American Fund for Environmental Cooperation (CEC)
NAFTA	North American Free Trade Agreement (Canada, USA, Mexico)
NAM	Non-Aligned Movement
NAP (1)	National Action Programme
NAP (2)	National Allocation Plan
NAPA	National Adaptation Programme of Action (UNFCCC)
NAPRI	North American Pollutant Release Inventory
NARBO	Asian River Basin Organization

NARS	National Agricultural Research System
NAS	National Academy of Sciences (USA)
NASA	National Aeronautics and Space Administration (USA)
NASDA	National Space Development Agency (Japan)
NASPD	National Association of State Park Directors
NATO	North Atlantic Treaty Organization
NAWMP	North American Waterfowl Management Plan
NBSAP	National Biodiversity Strategies and Action Plans
NC	National Committee
NC1	First National Communication
NCAR	National Center for Atmospheric Research (USA)
NCB	National Coordinating Body
NCC	Net Contributor Country
NCEA	National Center for Environmental Assessment
NCESD	National Council for Environment and Sustainable Development
NCSP (1)	National Communications Support Programme (GEF)
NCSP (2)	National Country Studies Programme (UNFCCC)
NDF	Nordic Development Fund
NEA	Nuclear Energy Agency (OECD)
NEAFF	Northeast Asian Forest Forum
NEAP	National Environmental Action Plan
NEDA	Netherlands Development Assistance, Ministry of Foreign Affairs
NEPAD	New Partnership for Africa's Development
NERC	Natural Environment Research Council (UK)
NEX	National Execution (GEF)
NFF	National Forest Foundation (USA)
NFI	National Forest Initiative
NFP (1)	National Forest Programme
NFP (2)	National Focal Point
NFPF	National Forest Programme Facility
NGA	Non-governmental Actor
NGIP	Non-Government Investment Program (IBRD)
NGLS	Non-Governmental Liaison Services (UN)
NGO	Non-Governmental Organization
NGOSC	NGO Steering Committee
NHP	National Historic Park
NIB	Nordic Investment Bank
NIEO	New International Economic Order
NIR	National Inventory Reporting
NIS	Newly Independent States

NMHS	National Meteorological and Hydrological Service (WMO)
NMSS	National Marine Sanctuary System (NOAA)
NMVOC	Non-Methane Volatile Organic Compounds
NNGO	National NGO
NNWS	Non-Nuclear-Weapon State
NOAA	National Oceanic and Atmospheric Administration
NORAD	Norwegian Agency for Development Cooperation
NOWPAP	North-West Pacific Action Plan
NPO	Non-Profit Organization
NPP	Net Primary Production
NPR	National Park Reserve
NPT	Treaty on the Non-Proliferation of Nuclear Weapons
NPV	Net present value
NRC	National Research Council (USA)
NRCAN	National Resources Canada
NRCC	National Research Council of Canada
NRDC	Natural Resources Defense Council
NRSE	New and Renewable Sources of Energy
NRTEE	National Round Table on the Environment and the Economy (Canada)
NSC	National Selection Committee
NSDC	National Sustainable Development Councils
NSF	National Science Foundation (USA)
NSSD	National Strategies for Sustainable Development
NTFP	Non-timber Forest Products
NWA	National Wildlife Area
NWFP	Non-wood Forest Products
NWP(1)	Nepal Water Partnership
NWP(2)	Netherlands Water Partnership
NWRI	National Water Research Institute (Canada)
NWS	Nuclear-Weapon State
NZAID	New Zealand Agency for International Development
NZODA	New Zealand Official Development Assistance

O

O&M	Operations and Maintenance
OA	Official Assistance

OAS	Organization of American States
OAS/USDE	OAS Unit for Sustainable Development and Environment
OAU	Organization of African Unity
OCA/PAC	Oceans and Coastal Areas Programme Activity Centre (UNEP)
OCC	Operational and Capital Costs
OCM	Ocean and Coastal Management
OD	Operational Directive (WB)
ODA (1)	Official Development Assistance
ODA (2)	Overseas Development Administration (UK)
ODAE	ODA Equity
ODAG	ODA Grant
ODAL	ODA Loan
ODI	Overseas Development Institute (UK)
ODP	Ozone Depleting Potential (also Ozone Destroying Potential)
ODPt	Ozone Depleting Potential in tons
ODS	Ozone Depleting Substance (USA)
OECD	Organization for Economic Cooperation and Development
OECS	Organization of Eastern Caribbean States
OED	Operations Evaluation Department (IBRD)
OeKB	*Oesterreichische Kontrollbank AG* (Austria)
OEWG	Open-ended Working Group
OFDA	Office of Foreign Disaster Assistance (USA)
OHP	Operational Hydrology Programme
OIC(1)	Organization of Islamic Countries
OIC(2)	Organization of the Islamic Conference
OILPOL	International Convention for the Prevention of Pollution of the Sea by Oil
OLADE	Latin American Energy Development Organization
OM	Operations Manual
ONGO	Operational NGO
OOFL	Other Official Flows of Non-concessional Lending by Multilateral Banks (OECD)
OP (1)	Operational Policy
OP (2)	Optional Protocol
OPCW	Organization for the Prohibition of Chemical Weapons
Op. Obj.	Operational Objective
OP/BP	Operation Plan/Business Plan (GEF-WB)
OPEC	Organization of Petroleum Exporting Countries
OPG	Operational Policy Guidelines
OPIC	Overseas Private Investment corporation (USA)

O

OPPRC	Oil Pollution Preparedness, Response and Coo-operation Convention
OPS (1)	Overall Performance Study (GEF)
OPS (2)	Office of Project Services (UNDP)
OPS (3)	Pan-American Health Organization (WHO)
OSCE	Organization for Security and Cooperation in Europe
OSPAR	Convention for the Protection of the Marine Environment of the Northeast Atlantic
OTF	Ozone Trust Fund
OTS	Organization for Tropical Studies

P

P	Preamble
P5	Permanent Five (Members of the UN Security Council)
P&Ms	Policies and Measures
PA	Participants Assembly (GEF)
PAC (1)	Interbureau Project Appraisal Committee
PAC (2)	Project Appraisal Committee
PACADIRH	Central American Water Resources Action Plan
PACD	Plan of Action to Combat Desertification
PAD	Project Appraisal Document (WB)
PADU	Protected Area Data Unit (WCMC)
PAG	Project Approval Group (UNEP)
PAGE	Pilot Analysis of Global Ecosystems (MA)
PAHO	Pan American Health Organization (WHO)
PAM (or P&Ms)	Policies and measures
PAME	Programme for the Protection of the Arctic Marine Environment
PAP	Priority Actions Programme
PAP-RAC	Regional Activity Centre for Priority Actions Programme (UNEP)
PARC	Performance Assessment Resource Centre
PAS	Protected Area Strategy
PBI	Programme Budget Implications
PBRs	Plant Breeders Rights
PC (1)	Plant Committee (CITES)
PC (2)	Permanent Council (OAS)
PCA	Permanent Court of Arbitration (UN)
PCBs	Polychlorinated Biphenyls

PC&I	Principles, Criteria, and Indicators
PCF	Prototype Carbon Fund (WB)
PCM	Project Cycle Management
PC-CP	From Potential (water) Conflicts to Cooperation Potential (UNESCO)
PDD	Project Design Document (UNFCCC)
PDF (1)	Project Preparation and Development Facility (GEF)
PDF (2)	Project Development Funds (GEF)
PD/GG	Participatory Development/Good Governance (OECD)
PDO	Project Development Objective
PEBLDS	Pan-European Biological and Landscape Diversity Strategy
PEDAS	Potentially Environmentally Detrimental Activities in Space
PERSGA	Regional Organization for the Conservation of the Environment of the Red Sea and Gulf of Aden
PETS	Public Expenditure Tracking survey (WB)
PFCs	Perfluorocarbons
PFII	United Nations Permanent Forum on Indigenous Issues
PG	Planning Group
PGRFA	Plant Genetic Resources for Food and Agriculture (CGRFA)
PHARE	EU's Assistance Programme in CE
PHVA	Population and Habitat Viability Assessment
PIC	Prior Informed Consent
PIC-INC	Intergovernmental Negotiating Committee for the Preparation of the Conference of Parties of the Rotterdam Convention for the Application of the Prior Informed Consent Procedure for Certain Hazardous Chemicals and Pesticides in International Trade
PID	Project Information Document (WB)
PIDP	Pacific Island Development Program
PIF	Pacific Island Forum
PINGO	Public Interest NGO
PIPR	Project Implementation Performance Report (GEF)
PIR	Project Implementation Review (GEF)
PM(1)	Prime Minister
PM(2)	Particulate Matter
PMS	Project Monitoring System (IDB)
PNA	Parties to the Nauru Agreement
PNG	Papua New Guinea
PO (1)	Private Organization
PO (2)	Peoples' Organization
PON	Program on Negotiations

POPs	Persistent Organic Pollutants (Stockholm Convention on ,)
PoWER	Partnership for Water Education and Research (UNESCO-IHE)
PP (1)	Project Purpose
PP (2)	Project Preparation (IDB)
PPA (1)	Project Preparation Advance (GEF)
PPA (2)	Project Preparation Assistance (GEF)
PPA (3)	Programme on Protected Areas (IUCN)
PPA (4)	Participatory Poverty Assessment
PPD	Proposed Project Document (IDB)
PPDOP	Participatory Process for the Definition of Options and Priorities
PPM	Production and Processing Methods
ppmv	Parts per million by volume
PPP (1)	Public-Private Partnership
PPP (2)	Purchasing Power Parities
PPR	Project Performance Review (GEF)
PRA	Participatory Rural Appraisal
PRC	People's Republic of China
PREC	Project Review and Evaluation Committee
PREM	Poverty Reduction and Economic Management Network (WB)
PRIF (1)	Pre-investment Financing
PRIF (2)	Pre-investment Facility
PRINCE	Program for Measuring Incremental Costs for the Environment (GEF)
PROBASE	Procedures for Accounting and Baselines for Projects under Joint Implementation and the Clean Development Mechanism.
PROFOR	Programme on Forests (UNDPWB)
ProDoc	Project Document
PRS	Poverty Reduction Strategy (WB)
PRSP	Poverty Reduction Strategy Paper (WB)
PRTR	Pollutant Release and Transfer Register
PSIA	Poverty and Social Impact Analysis (WB)
PSM	Public Sector Management
PTC	Program of Technical Cooperation (IDB)
PUB	Prediction in Ungauged Basins
PV	Photo Voltaic
PVO	Private Voluntary Organization
PVP	Plant Variety Protection
PWWA	Philippines Water Works Association
PWA	Portfolio of Water Actions (UNESCO)
PWP	Philippines Water Partnership
PWRI	Public Works Research Institute (Japan)

Q

QA/QC	Quality Assurance/Quality Control
QMS	Qualified Majority Voting System
QUELROs	Quantified Emission limitation and Reduction Objectives
QuNGO	Quasi-governmental NGO

R

RAC	Regional Activity Center
RAFI	Rural Advancement Fund International
RAIS	Regional Agricultural Information System
RAMSAR	Convention on Wetlands of International Importance Especially as Waterfowl Habitat
RAP (1)	Remedial Action Plan
RAP (2)	Regional Action Plan
RAP (3)	Resettlement Action Plan
RAP (4)	Rapid Assessment Project
RBC	River Basin Commission
RBD	River Basin District
RBO	River Basin Organization
RCs	Resource Centers
RCU	Regional Coordinating Unit
RCUWM	Regional Centre for Urban Water Management, Tehran, Iran (UNESCO)
R&D	Research and Development
RDB	Regional Development Bank
RDP (1)	Reconstruction Development Plan
RDP (2)	Regional Development Plan
REACH	Registration Evaluation, and Authorization of Chemicals
Rec	Recommendation (Document Identification)
REC	Regional Environmental Centre for Central and Eastern Europe
REEF	Renewable Energy and Energy Efficiency Fund (GEF)
RERеP	Regional Environmental Reconstruction Programme (GEF)
Res	Resolution (Document Identification)
Rev	Revision (Document Identification)

RFA	Recommendations for Action
RFC	Regional Forestry Commission (FAO)
RFP	Request for Proposal
RH	Reproductive Health
RIIA	The Royal Institute of International Affairs (UK)
RING	Alliance of Policy Research Organizations
RINGO	Research and Non-Governmental Organizations
RIO+10	The World Summit on Sustainable Development (Johannesburg 2002)
RIOs	Regional Indigenous Organizations
RIPANAP	Ibero-American Network of National Park Institutions and other Protected Areas
RIS	Ramsar Information System
RMUs	Removal Units (UNFCCC)
RO	Renewable Obligation
ROAP	Regional Office for Asia and the Pacific (UNEP)
ROC	Renewable Obligation Certificate
ROE	Regional Office for Europe (UNEP)
ROLAC	Regional Office for Latin America and the Caribbean (UNEP)
RONA	Regional Office for North America (UNEP)
ROPME	Regional Organization for the Protection of the Marine Environment
RPA	Research Priority Area
RPC	Regional Programming Committee
RPT95	Report Submitted in 1995
RR	Rules and Regulations (IMF)
RRA	Rapid Rural Appraisal
RRC.AP	Regional Resource Centre for Asia and Pacific
RSO	Research and Systematic Observation (UNFCCC)
RTAC	Regional Technical Advisory Committee (GWP)
RU (1)	Removal Unit (LULUCF)
RU (2)	Regional Unit of IBRD (AFR, Africa; EAP, East Asia and Pacific; ECA, Europe and Central Asia; LCR, Latin America and the Caribbean; MNA, Middle East and North Africa; SAR, South Asia)
RVP	Regional Vice President (WB)
RWP	Regional Water Partnership (GWP)

S

SAARC	South Asian Association for Regional Cooperation
SAC	Scientific Advisory Committee (UNEP)
SACE	*Sezione Speciale perl' Assicurazione del Credito all'Esportazione* (Italy)
SACEP	South Asian Cooperative Environment Programme
SADC	Southern African Development Community
SADCC	Southern African Development Coordination Conference
SAEFL	Swiss Agency for Environment, Forests and Landscape (BUWAL)
SAICM	Strategic Approach on International Chemicals Management
SAL	Structural Adjustment Loan
SAMTAC	South American Technical Advisory Committee (GWP)
SAP (1)	Strategic Action Plan (GEF)
SAP (2)	Scientific Assessment Panel
SAPTA	South Asian Preferential Trade Agreement
SAR (1)	Second Assessment Report (IPPC)
SAR (2)	Staff Appraisal Report (WB)
SARDC	South African Research and Documentation Center
SAREC	Department of Research Cooperation (Sida)
SARPN	South African Regional Poverty Network
SAR-TA	Staff Appraisal Report-Technical Annex (WB)
SAS	South Asian Seas Programme
SASTAC	South Asian Seas Technical Advisory Committee (GWP)
SATAC	Southern Africa Technical Advisory Committee (GWP)
SB	Subsidiary Body
SBD	Secretariat for Biological Diversity (UNEP)
SBI	Subsidiary Body for Implementation (UNFCCC)
SBSTA	Subsidiary Body for Scientific and Technological Advice (UNFCCC)
SBSTTA	Subsidiary Body on Scientific, Technical and Technological Advice (CBD)
SC (1)	Standing Committee
SC (2)	Steering Committee (GWP)
SC (3)	Sale of Children
SCAR	Scientific Committee on Antarctic Research
SCBD	Secretariat of the Convention on Biological Diversity
SCOPE	Scientific Committee on Problems of the Environment (ICSU)
SCOR	Scientific Committee on Oceanic Research
SCP	Socio-Cultural Profile

S

SD	Sustainable Development
SDC (1)	Swiss Development Corporation
SDC (2)	Solar Development Capital (GEF)
SDG	Sustainable Development Governance
SDR	Special Drawing Rights
SDV	Social Development Department (WB)
SEA(1)	Strategic Environmental Assessment
SEA(2)	Sustainable Enterprise Academy (Canada)
SEAGA	Socio-Economic and Gender Analysis
SEATAC	South East Asia Technical Advisory Committee (GWP)
SECO	*Secrétariat d'État á l'Économie* (Switzerland)
SEDAC	Socioeconomic Data and Applications Center (CIESIN)
SEED	Sustainable Energy and Environment Division (UNDP)
SEEDS	Sustainable Environment and Ecological Development Society
SEGA	Socio-Economic and Gender Analysis
SEI	Stockholm Environment Institute
SEQ	Standing Group on Emergency Questions (IEA)
SER.(1)	Series (OAS documents)
SER (2)	Society for Ecological Restoration International
SETAC	Society of Environmental Toxicology and Chemistry
SEWA	Self-Employed Women's Association (India)
SFM	Sustainable Forest Management (UNFF)
SG	Secretary General
SGP	Small Grants Programme
SHD	Sustainable Human Development
SHO	Self-help Organization
SHP	Small Hydropower
SHPO	Self-help Support Organization
SI	International System of Units
SICA	Central American Integration System
SICAP	Central American System of Protected Areas
SID	Society for International Development
SIDA (Sida)	Swedish International Development Agency
SIDB	Small Industries Development Bank
SIDS	Small Island Developing States
SIMDAS	Sustainable Integrated Management and Development of Arid and Semi-arid regions of Southern Africa
SINAP	National System of Protected Areas (Mexico)
SINDAS	Sustainable Integrated Management and Development of Arid and Semi-Arid Regions of Southern Africa (UNESCO)

SINGER	System-Wide Information Network on Genetic Resources
SIRG	Summit Implementation Review Group (Summit of the Americas Process)
SIWI	Stockholm International Water Institute
SLR	Side-looking Radar
SLT	Standing Group on Long-Term Cooperation (IEA)
SME	Small and Medium Enterprises
SME/SMI	Small and Medium-scale Enterprise/Industry
SMME	Small, Micro and Medium-sized Enterprises
SMO	Social Movement Organization
SMPR	Secretariat Managed Project Review (GEF)
SNA	System of National Accounts
SNV	Swedish Environment Protection Agency
SNW	Sustainable Northwest (USA)
SO	Strategic Objective
SOA	Summit of the Americas
SOD	Summary, Overview and Development Report (ICDA)
SOFO	State of the World's Forests Report
SOM	Standing Group on Oil Market (IEA)
SOPAC	South Pacific Applied Geoscience Commission
SPA	Specially Protected Areas and Wildlife (CAP)
SPAW	Protocol on Specially Protected Areas and Wildlife of the Cartagena Convention
SPC (1)	South Pacific Commission
SPC (2)	Secretariat for the Pacific Community
SPF	South Pacific Forum
SPFS	Special Programme for Food Security
SPM	Summary for Policy Makers (IPPC/Reviews)
SPMs	Sanitary and Phytosanitary Measures
SPS	Sanitary and Phytosanitary Agreement
SPOT	*Systeme Probatoire d'Observation de la Terre*
SPREP	South Pacific Regional Environmental Programme (UNEP)
SPS	Sanitary and Phytosanitary Agreement (WTO)
SRAP	Sub-Regional Action Plan
SRES	Special Report on Emissions Scenarios
SRFC	Sub-Regional Fisheries Commission
SRH	Secretariat for Water Resources (Brazil)
SRI	Socially Responsible Investment
SRLFC	Special Report on Land Use, Land Use Change and Forestry and Carbon Sinks (IPCC)

S

SS	Special Session
SSC	Species Survival Commission (IUCN)
SSP	Social Policy Programme (IUCN)
SSWP	Small Scale Water Provider (ADB)
STAP	Scientific and Technical Advisory Panel (GEF)
START (1)	Global Change Systems for Analysis, Research and Training
START (2)	Strategic Arms Reduction Treaty
STATE	US State Department
STC	Short-term Consultant (WB)
STI	Sexually Transmitted Infection
STRP	Scientific and Technical Review Panel (Ramsar)
SUI	Sustainable Use Team (IUCN)
SWCC	Second World Climate Conference
SWG	Sub Working Group
SWR	Sub-Committee on Water Resources (ACC)
SWS	Safe Water Systems

T

TA	Technical Assistance
TAC (1)	Total Allowable Catch
TAC (2)	Technical Advisory Committee
TAG	Technical Advisory Group
TAR	Third Assessment Report (IPPC)
TBD	Tropical Biodiversity (CHM)
TBT	Technical Barrier to Trade
TCAs	Transboundary Conservation Areas
TCBO	Training, Capacity Building, and Outreach
tCe	Tons of carbon equivalent
TCP	Technical Cooperation Programme (FAO)
TDA (1)	Transboundary Diagnostic analysis (GEF)
TDA (2)	US Trade and Development Agency
TE	Total Expenditures
TEAP	Technology and Economic Assessment Panel
TEC	Technical Committee (GWP)
TEK	Traditional Ecological Knowledge
TEMS	Terrestrial Ecosystem Monitoring Sites (Database)
TERI(1)	Terrestrial Ecosystem Resource Inventory (CI)

TERI(2)	The Energy and Resources Institute
TERM	Transport and Environment Reporting Mechanism
TFAP	Tropical Forestry Action Plan
TFCA	Tropical Forest Conservation Act (USA)
TFCP	Tropical Forest Canopy Programme
TFDT	Task Force on Destruction Technologies
TFF	Tropical Forestry Foundation
TFI	Task Force for National Greenhouse Inventories
TFRK	Traditional Forest-Related Knowledge (CBD)
TGCIA	Task Group on Scenarios for Climate and Impact Assessment (IPCC)
TI	Transparency International
TKBD	Traditional Knowledge and Biological Diversity
TM	Task Manager
TMB	Trust Management Board
TNC (1)	The Nature Conservancy
TNC (2)	Transnational Corporation
TOC	Technological Options Committee
TOE	Tons of Oil Equivalent
TOR	Terms of Reference
TPR	Tripartite Review (GEF)
TRAFFIC	Trade Record Analysis of Flora and Fauna in Commerce
TREES	Tropical Ecosystem Environment Observations by Satellite (EEC)
TRI	Toxic Release Inventory
TRIMs	Trade-related Investment Measures
TRIPs	Agreement on Trade-related Aspects of Intellectual Property Rights (IU)
TRN	Taiga Rescue Network
TRP	Technical Review Panel (GEF)
TRR	Traditional Resource Rights
TS(1)	Technical Summary
TS(2)	Technical Secretariat
TSU	Technical Support Unit (IPCC)
TTCLEAR	Technology Information Clearing House
TTF	Thematic Trust Fund
TWAS	Third World Academy of Sciences
TWUWS	Transportation, Water and Urban Development Department (UNDP-WB)
TWWF	Third World Water Forum

U

U

UDHR	Universal Declaration of Human Rights
UESNET	Urban Environmental Sanitation Network (GWP)
UIE	UNESCO Institute for Education (Hamburg, Germany)
UIS	UNESCO Institute for statistics (Montreal, Canada) (UNESCO)
UIUC	University of Illinois at Urbana-Champaign
UN	United Nations
UNACABQ	UN Advisory Committee on Administrative and Budgetary Questions
UNACC/SCWR	United Nations Administrative Committee on Coordination Subcommittee on Water Resources
UNAIDS	United Nations Fund for AIDS
UNCBD	United Nations Secretariat of the Convention on Biological Diversity
UNCCD	United Nations Convention to Combat Desertification
UNCEB	United Nations Chief Executives Board
UNCED	United Nations Conference on Environment and Development
UNCEF	United Nations Capital Development Fund
UNCHE	United Nations Conference on Human Environment
UNCHR	United Nations Commission on Human Rights
UNCHS	United Nations Commission on Human Settlements
UNCLOS	United Nations Convention on the Law of the Sea
UNCPD	United Nations Commission on Population and Development
UNCRC	United Nations Convention on the Rights of the Child
UNCRD	United Nations Centre for Regional Development
UNCTAD	United Nations Conference on Trade and Development
UNCUEA	United Nations Centre for Urgent Environmental Assistance
UNDAC	United Nations Disaster Assessment and Coordination Team
UNDAF	United Nations Development Assistance Framework
UNDCP	United Nations Drug Control Program
UNDESA	United Nations Department of Economic and Social Affairs
UNDHR	United Nations Declaration on Human Rights
UNDP	United Nations Development Programme
UNDP/IW-LEARN	International Waters Learning Exchange and Resource Network
UNDPRF	United Nations Development Programme Revolving Fund
UNDRO	United Nations Office of Disaster Relief
UNECA	United Nations Economic Commission for Africa

UNECE	United Nations Economic Commission for Europe
UNECLAC	United Nations Economic Commission for Latin America and the Caribbean
UNEO	United Nations Environment Organization
UNEP	United Nations Environment Programme
UNEP-CAR/RCU	Caribbean Regional Coordinating Unit of the Cartagena Convention (UNEP)
UNEP-CEP	United Nations Environment Programme Caribbean Environment Programme
UNEP-EAS/RCU	United Nations Environment Programme East Asia Seas Regional Coordinating Unit
UNEP-ENRIN	United Nations Environment Programme Environment and Natural Resource Information Group
UNEP-GC	United Nations Environment Programme Governing Council
UNEP-GPA	United Nations Environment Programme Global Programme of Action or the Protection of the Marine Environment from Land-based Activities
UNEP-GRID	United Nations Environment Programme Global Resources Information Database
UNEP-HEM	United Nations Environment Programme Harmonization of Environmental Measurement
UNEP-IETC	United Nations Environment Programme International Environment Technology Centre
UNEP-MEDU	United Nations Environment Programme Coordinating Unit of the Mediterranean Action Plan
UNEP-OS	United Nations Environment Programme Ozone Secretariat
UNEP-RCUs	United Nations Environment Programme Regional Coordinating Units of the Regional Seas Programme
UNEP-ROA	United Nations Environment Programme Regional Office for Africa
UNEP-ROAP	United Nations Environment Programme Regional Office for Asia and the Pacific
UNEP-ROE	United Nations Environment Programme Regional Office for Europe
UNEP-ROLAC	United Nations Environment Programme Regional Office for Latin America and the Caribbean
UNEP-RONA	United Nations Environment Programme Regional Office for North America
UNEP-ROWA	United Nations Environment Programme Regional Office for West Asia

UNEP-SPAW	Specially Protected Areas and Wildlife Protocol of the Cartagena Convention
UNEP-WCMC	UNEP World Conservation Monitoring Center
UNESCAP	United Nations Economic and Social Commission for Asia and the Pacific
UNESCO	United Nations Educational, Scientific, and Cultural Organization
UNESCO-IHE	UNESCO Institute for Water Education (2003, Holland)
UNESCO-IHP	UNESCO International Hydrological Programme
UNEVOC	International Centre for technical and Vocational Education and Training (Bonn, Germany) (UNESCO)
UNF	United Nations Foundation
UNFAO	United Nations Food and Agricultural Organization
UNFCCC	United Nations Framework Convention on Climate Change
UNFF	United Nations Forum on Forests
UNFIP	United Nations Fund for International Partnerships
UNFPA	United Nations Fund for Population Activities
UNG/WG	United Nations Geographical Information Working Group
UNGA	United Nations General Assembly
UNGASS	United Nations General Assembly Special Session (i.e. UN General Assembly Special Session to Assess the Earth Summit)
UNGIWG	United Nations Geographic Information Working Group
UNHCR	United Nations High Commissioner for Refugees
UN-Habitat	United Nations Programme for Human Settlements
UNIA	United Nations Implementing Agreement
UNICEF	United Nations Children's Fund
UNICJ	United Nations International Court of Justice
UNIDCP	United Nations International Drug Control Programme
UNIDO	United Nations Industrial Development Organization
UNIDROIT	International Institute for the Unification of Private Law
UNIFEM	United Nations Development Fund for Women
UNISDR	United Nations International Strategy for Disaster Reduction
UNITAR	United Nations Institute for Training and Research
UNMOVIC	United Nations Monitoring, Verification and Inspection Commission
UNON	United Nations Office at Nairobi
UNMDGs	United Nations Millenium Development Goals
UNOPS	United Nations Office for Project Services
UNSD	United Nations Statistics Division
UNSIA	United Nations Special Initiative for Africa
UNSO	United Nations Sudano-Saheilian Office

UNSTAT	United Nations Statistical Division
UNT	United Nations Treaty Collection
UNU	United Nations University
UNU-IAS	United Nations University Institute for Advanced Studies
UNU-INWEH	United Nations University International Network on Water, Environment, and Health
UNWater	United Nations Committee on Fresh Water
UNWWAP	United Nations World Water Action Programme
UPOV	International Union for the Protection of New Varieties of Plants
UPTW	University Partnership for Transboundary Water (UNESCO)
URF	Universal Reporting Format
USA	United States of America
USACE	US Army Corps of Engineers
USAID	United States Agency for International Development
USCCSP	United States Climate Change Science Program
USDE	Unit of Sustainable Development and Environment (OAS)
USEPA	United States Environmental Protection Agency
USFS	United States Department of Agriculture Forest Service
USFWS	United States Fish and Wildlife Service
USGS	United States Geological Survey
UTF	Unilateral Trust Fund (FAO)

V

V&A	Vulnerability and Adaptation Assessment
VA	Voluntary Agreement (IEA)
VASTRA	Swedish Water Management Research Programme
VAT	Value Added Tax
VCLT	Vienna Convention on the Law of Treaties between States and International Organizations or between International Organizations (1986)
VEA	Voluntary Environmental Agreement
Vienna/ Montreal	Vienna Convention for the protection of the Ozone layer and its Montreal Protocol on Substances that Deplete the Ozon Layer
VITA	Volunteers in Technical Assistance
VMS	Vessel Monitoring System
VP	Vice President
VTPI	Victoria Transport Policy Institute (Canada)

W

WACLAC	World Association of Cities and Local Authorities Coordination
WANA	West Asia and North Africa
WARDA	West Africa Rice Development Association
WARFSA	Water Research Foundation for Southern Africa
WASER	World Association for Sedimentation and Erosion Reseach
WATAC	West Africa Technical Advisory Committee (GWP)
WB	World Bank
WBCSD	World Business Council for Sustainable Development
WBER	World Bank Economic Review
WBG	World Bank Group
WBGU	German Advisory Council on Global Change
WBI	World Bank Institute
WCA	Water Conservation in Agriculture (GWP)
WCAR	World Conference against Racism
WCASP	World Climate Application and Service Programme
WCC	World Conservation Congress
WCD	World Commission on Dams
WCF	Water Cooperation Facility (UNESCO, WWC)
WCDP	World Climate Data Programme
WCED	World Commission on Environment and Development (The Brundtland Commission)
WCFSD	World Commission on Forests and Sustainable Development
WCIP	World Climate Impact Study Programme (UNEP)
WCMC	World Conservation Monitoring Centre
WCO	World Customs Organization
WCP	World Climate Programme
WCPA	World Commission on Protected Areas (IUCN)
WCRP	World Climate Research Programme
WCS (1)	World Conservation Strategy
WCS (2)	Wildlife Conservation Society
WCSIW	Wider Caribbean Initiative for Ship-Generated Waste
WDC	World Data Centre (ICSU)
WDCS	Whale and Dolphin Conservation Society
WDI	World Development Indicators
WDR	World Development Report
WE	Western Europe
WEA	World Energy Assessment

WEC (1)	World Energy Council
WEC (2)	World Environment Center
WEC (3)	World Environment Capacity
WEDO	Women's Environment and Development Organisation
WEF	World Economic Forum
WEHAB	Water, Energy, Health, Agriculture and Biodiversity objectives articulated at WSSD
WEI	World Environment Institute
WEO (1)	World Environment Organization
WEO (2)	World Energy Outlook
WEOG	Western Europe and Others Group
WES	Water, Environment and Sanitation Programme
WET	Water Education for Teachers (USA)
W-E-T	Water-Education-Training (UNESCO)
WFC	World Food Council
WFD	Water Framework Directive (EU)
WFE	Water, Food and Environment
WFED	World Foundation for Environment and Development
WFP	World Food Programme
WFS	World Food Summit of 1996
WG	Working Group
WGEM	Ad Hoc Working Group on Environmental Monitoring of the UNECE
WGMS	World Glacier Monitoring Service
WHA	World Health Assembly
WHC	World Heritage Convention
WHO	World Health Organization
WHYCOS	World Hydrological Observing System
WID	Women in Development
WIDECAST	Wider Caribbean Sea Turtle Conservation Network
WIPO	World Intellectual Property Organization
WIS	Water Information Summit (IWRN)
WMO	World Meteorological Organization
WOCAR	World Conference against Racism
World Bank	International Bank for Reconstruction and Development (IBRID) and the International Development Association (IDA)
WPC	World Parks Congress
WPI	Water Poverty Index
WQMP	Water Quality Management Programme (GWP)
WRI	World Resources Institute

WSC	Water Sector Committee (ADB)
WSCU	Water Sector Coordination Unit
WSF	World Solidarity Fund (UN)
WSIS	World Summit on the Information Society
WSP (1)	Water Safety Plans
WSP (2)	Water and Sanitation Programme (GWP)
WSSCC	Water Supply and Sanitation Collaborative Council
WSSD	World Summit on Sustainable Development
WTA	World Trade Agreement
WTO(1)	World Trade Organization
WTO(2)	World Tourist Organization
WUP	Water Utilities Partnerships in Africa ((GWP)
WW2BW	From white water to blue water
WWAP	World Water Assessment Programme (UNESCO)
WWA-WAU	World Water Council-Water Action Unit
WWC	World Water Council
WWDR	World Water Development Report
WWF (1)	World Wide Fund for Nature (World Wildlife Fund, World Wide Fund International)
WWF (2)	World Water Forum (WWC)
WWV	World Water Vision
WWW	World Weather Watch (WMO)

Z

ZIL	Swiss Centre for International Agriculture
ZNWP	Zimbabwe National Water Partnership
ZOPP	Objectives-Oriented Project Planning
ZWA	Zero Waste Alliance

Bibliography

Abbott, Kenneth W. and Duncan Snidal. (2000). Hard and Soft Law in International Governance. *International Organization*, 2000, 54.3: 421-56.

Albert, A. (ed.) (1995). *Chaos and Society*, Amsterdam: IOS Press.

Arthur, W. Brian. (1993). Why Do Things Become More Complex? *Scientific American*, May: 144.

Asian Development Bank. (1986). *Environmental Guidelines for the Development of Ports and Harbours*, Manila: ADB.

Axelrod, Robert. (1997). *The Complexity of Cooperation*, Princeton: Princeton University Press.

Benedick, Richard Elliot. (1993). Perspectives of a Negotiation Practitioner. In Sjostedt, Gunnar (ed.) *International Environmental Negotiation*, London: Sage Publications, 219-43.

Bharati, Veda. (2001). *Yoga Sutras of Patanjali with the Exposition of Vyasa: A Translation and Commentary*. Motilal Banarsidas Publishers, Delhi.

Biermann, Frank, Rainier Brohm and Klaus Dingwerth (eds.). (2002). *Proceedings of the 2001 Berlin Conference on the Human Dimensions of Global Environmental Change: Global Environmental Change and the Nation State*. PIK Report No. 80. Potsdam: Potsdam Institute for Climate Impact Research.

Bigg, Tom. (2003). The World Summit on Sustainable Development: Was it worthwhile? Accessed at http://www.iied.org/wwsd/publish.html on 15 October 2003.

Birnie, Patricia and Alan E. Boyle. (1992). *International Law & the Environment*, Oxford: Clarendon Press.

Boot, Max. (2002). The Big Enchilada: American Hegemony will be Expensive. *The International Herald Tribune*, 15 October.

Botkin, Daniel B. (1990). *Discordant Harmony: A New Ecology for the Twenty-first Century*, Oxford: Oxford University Press.

Brown, Becky J., Mark E. Hanson, Diana M. Liverman, and Robert W. Merideth Jr. (1987). Global Sustainability: Toward Definition. *Environmental Management*, 11.6: 713-19.

Caldwell, Lynton. (1990). *Between Two Worlds: Science, the Environmental Movement, and Policy Choice*, Cambridge: Cambridge University Press.

Cambel, A. B. (1993). *Applied Chaos Theory: A Paradigm for Complexity*, San Diego: Academic Press, Inc.

Centre for Global Development and the Carnegie Endowment for International Peace. (2003). Ranking the Rich. *Foreign Policy*, May June: 56-66.

Chayes, Abram and Antonia H. Chayes. (1991). Adjustment and Compliance Processes in International Regulatory Regimes, In: Tuchman Mathews, Jessica (ed.). *Preserving the Global Environment: The Challenge of Shared Leadership*, New York and London: W. W. Norton and Company. 280-308.

Columbia University. (N.D.) CIESIN Data Base. Accessed at http://sedac.ciesin.columbia.edu.

Cooper, Andrew F., John English and Ramesh Thakur. (2002). Enhancing Global Government: Towards a New Diplomacy, Tokyo: UNU Press.

Corell, Elizabeth and Michele M. Betsell. (2001). A Comparative Look at NGO Influence in Intergovernmental Environmental Negotiations: Desertification and Climate Change. *Global Environmental Politics*, 1.4: 86-107.

Daly, Herman E. (1996). *Beyond Growth: The Economics of Sustainable Development*, Boston: Beacon Press.

Domoto, Akiko. (2001). "International Environmental Governance: Its Impact on Social and Human Development." *Inter-linkages*, World Summit for Sustainable Development. United Nations University Centre. 3-4 September.

Drezner, Daniel W. (2002).Bargaining, Enforcement, and Multilateral Sanctions: When is Cooperation Counterproductive? *International Organization*, 54.1: 73-102.

Dupont, Christophe. (1993). The Rhine: A Study of Inland Water Negotiations. In: Sjostedt, Gunnar (ed.) *International Environmental Negotiation*, London: Sage Publications. 135-48.

Fauth, John E. (1997). Working toward Operational Definitions in Ecology: Putting the System back into Ecosystem. *Bulletin of the Ecological Society of America*, 78.4: 295.

Flavin, Christopher et al. (eds.). (2002). *State of the World 2002-Progress Towards a Sustainable Society*, London: Worldwatch Institute and W.W. Norton and Company.

French, Hillary. (2001). *Vanishing Borders*, London: Worldwatch Institute and W.W. Norton and Company.

French, Hillary. (2002). Reshaping Global Governance, *State of the World 2002. Progress Towards a Sustainable Society*, Christopher Flavin et al. (eds.), London: Worldwatch Institute and W.W. Norton and Company, 174-98.

Fridtjof Nansen Institute. (2001). *Yearbook of International Cooperation on Environment and Development*. London: Earthscan

Gallopin, Gilberto C. (1981). Human Systems: Needs, Requirements, Environments and Quality of Life. In G.E. Lasker (ed.) *Applied Systems and Cybernetics. Vol. 1. The Quality of Life: Systems Approaches*, Oxford: Pergamon Press.

Gallopin, Gilberto C. (1981). The Abstract Concept of Environment. *Int. J. General Systems*, 7: 139-49.

Goulet, Denis. (1986). Three Rationalities in Development Decision-making. *World Development*, 14.2: 301-17.

Gupta, Joyeeta. (2002). *"On behalf of My Delegation..." A Survival Guide for Developing Country Climate Negotiators*. Climate change Knowledge Network/Center for Sustainable Development in the Americas. Accessed on July 22, 2002 at
http://www.unitar.org/cctrain/Survival%20Negotiators%(nAIP)www/index.htm.

Haas, Peter. (1992). Introduction. *Epistemic Communities and International Policy Coordination*, International Organization, 46(1):1-35.

Hawley Amos H. (1986). *Human Ecology: A Theoretical Essay*, Chicago: The University of Chicago Press.

Hempel, Lamont. (1996). *Environmental Governance: the Global Challenge*, Washington D.C.: Island Press.

Hewson, Martin and Timothy J. Sinclair (eds.). (1999). *Approaches to Global Governance Theory*, Albany, NY: State University of New York Press.

Hollings, C. S. (1973). Resilience and Stability of Ecological Systems. *Annual Review of Ecology and Systematics*, 4:1-24.

Hunter, David, James Salzman, and Durwood Zaelke. (1998). *International Environmental Law and Policy*, New York: Foundation Press.

IISD. (2001). Summary Report from the UNEP Expert Consultations on International Environmental Governance, 2nd Round Table, 29 May 2001. *IISD Linkages* 53.1.

Ingerson, Alice E. (2002). A Critical User's Guide to "Ecosystem" and Related Concepts in Ecology. Institute for Cultural Landscape Studies, the Arnold Arboretum of Harvard University. Accessed 2003, at
http://www.icls.harvard.edu/ecology/ecology.html

IPCC. (2001). Third Assessment Report. Intergovernmental Panel on Climate Change. WHO and UNEP. Accessed on July 16, 2003 at http://www.ipcc.ch.

IUCN, UNEP and WWF. (1980). *World Conservation Strategy: Living Resource Conservation for Sustainable Development*, Gland: IUCN.

Jakobson, Max. (2002). Preemption: Shades of Roosevelt and Stalin. *The International Herald Tribune*, 17 October.

Kagan, Robert. (2002). Power and Weakness, *Policy Review on Line*. Accessed 15 August 2002 at http://www.policyreview.org/JUN02/kagan.html.

Kay, James J. (1991). A Non-equilibrium Thermodynamic Framework for Discussing Ecosystem Integrity. *Environmental Management*, 15.4: 483-95.

Keating, Michael. (1993). *The Earth Summit's Agenda for Change: A plain language version of Agenda 21 and the other Rio Agreements*, Geneva: Centre for Our Common Future.

Kellert, Stephen. (1995). When is the Economy Not like the Weather? The Problem of Extending Chaos Theory to the Social Sciences. In: Albert, A. (ed.) *Chaos and Society*, Amsterdam: IOS press.

Kremenyuk, Victor and Winfried Lang. (1993). The Political, Diplomatic, and Legal Background. In Sjostedt, Gunnar (ed.) *International Environmental Negotiation*, London: Sage Publications. 3-16.

Lackey, Robert T. (2001). Values, Policy, and Ecosystem Health. *BioScience*, 51.6: 437-43.

Lean, Geoffrey. (2002). U.N. Creates Watchdog Group in Lieu of Future Summits, *London Independent*, 8 September.

Lipschutz, Ronnie D. (1999). From Local Knowledge and Practice to Global Environmental Governance. In: Hewson, Martin and Timothy J. Sinclair (eds.). *Approaches to Global Governance Theory*, Albany, NY: State University of New York Press. 259-83.

Lipschutz, Ronnie D. (2003). *Global Environmental Politics from the Ground Up*. Accessed at http://ic.ucsc.edu/~rlipsch/pol174/syllabus.html on July 15, 2003.

Lorenz, Edward N. (1993). *The Essence of Chaos*, Seattle: University of Washington Press.

Lugo, Ariel E. (1991). Stress and Ecosystems, In: J. H. Thorp and J. W. Gibbons (eds.) *Energy and Environmental Stress in Aquatic Systems*, DOE Symposium Series (1978).

Maerz, John C. (1994) Ecosystem Management: A Summary of the Ecosystem Management Roundtable of 19 July 1993. *Bulletin of the Ecological Society of America*, 75.2: 93-5.

Manners, Ian R., and Marvin W. Mikesell (eds.). (1974). *Perspectives on Environment*, Washington D.C.: Association of American Geographers.

Marín, Victor. (1997). General System Theory and the Ecosystem Concept" *Bulletin of the Ecological Society of America*, 78.1: 102-103.

Miller, Kenton R. (1972). International Cooperation Programs Which Relate to National Parks and Wildland Management, *Planning National Parks for Ecodevelopment: Methods and Cases from Latin America*, Washington D.C. Peace Corps Information Collection and Exchange. Accessed at: http://mng-unex1.marasconewton.com/peacecorps/documents/ROO73/r0073/r0073e/r0073e00.htm.

Mitchell, Ronald B. (2002a). International Environment, In: Risse, Thomas, Beth Simmons, and Walter Carlsnaes (eds.) Handbook of International Relations, London: Sage Publications.

Mitchell, Ronald B. (2002b). Of Course International Institutions Matter: But When and How? In: Biermann, Frank, Rainier Brohm and Klaus Dingwerth (eds.) *Proceedings of the 2001 Berlin Conference on the Human Dimensions of Global Environmental Change: Global Environmental Change and the Nation State*, PIK Report No. 80. Potsdam: Potsdam Institute for Climate Impact Research. 16-25.

Mitchell, Ronald B. (2003). International Environmental Agreements Defined." Accessed at http://www.uoregon.edu/~rmitchel/IEA/overview/definitions/htm, on 15 October.

Najan, Adil. (1995). An Environmental Negotiation Strategy for the South. *International Environmental Affairs*, 7.3: 249-87.

Nonaka, Ikujiro, and Hirotaka Takeuchi. (1995). *The Knowledge-Creating Company; How Japanese Companies Create the Dynamics of Innovation*, Oxford University Press, New York.

OAS. (1987). *Minimum Conflict: Guidelines for Planning the Use of American Humid Tropic Environments*, Washington D. C.: Organization of American States.

OAS. (2001). *Inter-American Strategy for Public Participation in Decision-Making for Sustainable Development*, Washington, D. C.: General Secretariat, Organization of American States.

Pannell, David J. and Steven Schilizzi. (1997). Sustainable Agriculture: A Question of Ecology, Equity, Economic Efficiency or Expedience? *Sustainability and Economics in Agriculture,* SEA Working Paper 97/1. GRDC Project, University of Western Australia.

Pavard, Bernard and Julie Dugale. (2003). An Introduction to Complexity in Social Science. Accessed July 15 at http://www.irit.fr/cosi/training/Complexity-tutorial/htm.

Pezzey, J. (1992). *Sustainable Development Concepts: An Economic Analysis*, World Bank Environment Paper No. 2. Washington, D.C.: The World Bank.

Pomeroy, Robin. (2002). Earth Summiteers Cast Doubt on Future World Meets. *Reuters News Publication Service*, 4 September.

Ramsundersingh, Atem. (2003). *Introducing Creative Learning at UNESCO-IHE*, Proceedings of the Conference on Water Education and Capacity Building (WECB), Third World Water Forum, Kyoto.

Risse, Thomas, Beth Simmons, and Walter Carlsnaes (eds.). (2002). *Handbook of International Relations*, London: Sage Publications.

Rubin, Jeffery Z. (1993). Third Party Roles: Mediation in International Environmental Disputes. In: Sjostedt, Gunnar (ed.) *International Environmental Negotiation*, London: Sage Publications. 275-90.

Saarinen, Thomas F. (1974). Environmental Perception. In: Ian R. Manners and Marvin W. Mikesell (eds.) *Perspectives on Environment*, Washington D.C.: Association of American Geographers. 252-89.

Saunier, Richard E. (1999). *Perceptions of Sustainability: A Framework for the 21st Century*. Trends for a Common Future 6, Washington, D. C.: CIDI/Organization of American States.

Sebenius, James K. (1993). The Law of the Sea Conference: Lessons for Negotiations to Control Global Warming. In: Sjostedt, Gunnar (ed.) International Environmental Negotiation, London: Sage Publications. 189-215.

SER. (2003). Global Rationale for Ecological Restoration. IUCN-CEM 2nd Ecosystem Restoration Working Group Meeting, 2-5 March 2003. Taman Negara, Malaysia.

Sharma, Anju, Richard Mahapatra and Clifford Polycarp. (2002). Dialogue of the Deaf, *Down to Earth*, Delhi: Centre for Science and Environment. 25-33.

Sjostedt, Gunnar (ed.). (1993). *International Environmental Negotiation*, London: Sage Publications.

Thorp, J. H. and J. W. Gibbons (eds.). (1978). *Energy and Environmental Stress in Aquatic Systems*. DOE Symposium Series.

Trzyna, T., Margold, E. and J.K. Osborn. (1996). *World Directory of Environmental Organizations*. California Institute of Public Affairs, Sacramento, CA. Accessed at http://www/cipahq/prg/landmarks.htm.

Tuchman Mathews, Jessica (ed.). (1991). *Preserving the Global Environment: The Challenge of Shared Leadership*, New York and London: W. W. Norton and Company.

UNCHS. (2002). *Cities in a Globalizing World: Global Report on Human Settlements 2001*, London: Earthscan Publications Ltd.

UNDESA/DSD. (2003). Partnerships for Sustainable Development. Accessed at wttp://www.un.org/esa/sustdev/partnerships/partnerships.htm on 13 November 2003.

UNDP. (2003). *Compendium of Environmental Projects*. (New York: United Nations Development Program. 1996). Accessed on 6 December 2003 at http://www.undp.org.vn/undp/docs/1996/envprojs/toc.htm.

UNECE. (2002). *Introducing the Aårhus Convention*. Accessed on 15 August 2001 at http://www.buwal.ch/inter/e/ea_zugan.htm.

UNEP. (1993). *Global Environment Outlook 3*, London: Earthscan Publications Ltd. 2002.

UNEP. (1997). Compendium of Legislative Authority 1992-1997, Oxford: United Nations Environment Programme/Express Litho Service.

United Nations Food and Agriculture Organization. (1997). *Food Production: the Critical role Role of Water*. Technical background document 7. World Food Summit, 13-17 November 1996. Rome.

UNGA. (1982). *World Charter for Nature*, UNGA RES 37/7.

United Nations. (n.d.). United Nations Treaty Data Base. Accessed at http://untreaty.un.org.

United Nations Millennium Development Goals. (2000). Accessed at http://www.un/org/millenniumgoals/.

Waldrop, M. Mitchell. (1993). *Complexity: The Emerging Science at the Edge of Order and Chaos*, New York. Touchstone.

Wali, Mojan. (1995). eco Vocabulary: A glossary of our times *Bulletin of the Ecological Society of America*, Vol 76: 106-11.

Weeks, Dudley. (1992). *Eight Essential Steps to Conflict Resolution*, New York: G.P. Putnam's Sons.

Wenger, Etienne. (1998). *Communities of Practice; Learning, Meaning and Identity*, Cambridge University Press, Cambridge.

World Commission on Environment and Development. (1987). *Our Common Future*, Oxford: Oxford University Press.

World Resources Institute. (2003). WRI Expresses disappointment over many WSSD Outcomes. Accessed at http://newsroom.wri.org on 6 October 2003.

WRI, IUCN, UNEP. (1992). *Global Biodiversity Strategy*, Washington, D.C.: World Resources Institute.

Wyant, James G., Richard A. Meganck, and Sam H. Ham. (1995). The need for an environmental restoration decision framework. *Ecological Engineering*, 5:417-20.

Young, Oran R. (ed). (1997). *Global Environmental Accord: Strategies for Sustainability and Institutional Innovations*, Cambridge: MIT Press.

Young, Oran R. (1997). Global Governance: Drawing Insights from the Environmental Experience, In: Young, Oran R. (ed.) *Global Environmental Accord: Strategies for Sustainability and Institutional Innovations*, Cambridge: MIT Press.

Zartman, William I. (1993). Lessons for Analysis and Practice In: Sjostedt, Gunnar (ed.) *International Environmental Negotiation*, London: Sage Publications, 262-74.

Appendices

Appendix 1

Principles and Values of Global Environmental Governance

Most GEG meetings, conferences, and summits such as UNCED or WSSD produce a final product, normally divided into a Declaration of Principles and a Plan of Action (POA). The Declaration of Principles, which is generally signed by the Ministers or Heads of State, is a broad document which provides the context for understanding the detailed initiatives or substantive mandates elaborated in the plan of action. These collected principles now help "guide" GEG and although many of them appear in declarations and are, therefore, soft law, others, because they are essential parts of universally ratified accords that have entered into force, have taken on the stature of international law. We list a few of the more important here.

Although not Principles or Values in a strict sense, the Millennium Development Goals (MDGs) are included in this appendix because for the first time in history all 191 UN Member States agreed to a concerted effort to address the fundamental issues affecting sustainable development. As a test of political will to build strong partnerships at all levels, the MDGs place responsibilities on both rich and poor nations. Rich nations are specifically called on to relieve debt, increase aid and give poor countries fair access to their markets and their technology. Developing countries have the responsibility to undertake policy reforms and strengthen governance. The Goals, which will guide both technical assistance and investment in developing countries for the next decade, offer the word a means to accelerate the pace of development and to actually measure the results.

United Nations Universal Declaration of Human Rights (1948)

1. All human beings are born free and equal in dignity and rights. They are endowed with reason and conscience and should act towards one another in a spirit of brotherhood.
2. Furthermore, no distinction shall be made on the basis of the political, jurisdictional or international status of the country or territory to which a person belongs, whether it be independent, trust, non-self-governing or under any other limitation of sovereignty.
3. Everyone has the right to life, liberty and security of person.
4. No one shall be held in slavery or servitude; slavery and the slave trade shall be prohibited in all their forms.
5. No one shall be subjected to torture or to cruel, inhuman or degrading treatment or punishment.
6. Everyone has the right to recognition everywhere as a person before the law.
7. All are equal before the law and are entitled without any discrimination to equal protection of the law. All are entitled to equal protection against any discrimination in violation of the Declaration and against any incitement to such discrimination.
8. Everyone has the right to an effective remedy by the competent national tribunals for acts violating the fundamental rights granted him by the constitution or by law.
9. No one shall be subjected to arbitrary arrest, detention or exile.
10. Everyone is entitled in full equality to a fair and public hearing by an independent and impartial tribunal, in the determination of his rights and obligations and of any criminal charge against him.
11. a. Everyone charged with a penal offense has the right to be presumed innocent until proved guilty according to law in a public trial at which he has had all the guarantees necessary for his defense.
    b. No one shall be held guilty of any penal offense on account of any act or omission which did not constitute a penal offense, under national or international law, at the time it was committed. Nor shall a heavier penalty be imposed than the one that was applicable at the time the penal offense was committed.
12. No one shall be subjected to arbitrary interference with his privacy, family, home or correspondence, nor to attacks upon his honor and reputation. Everyone has the right to the protection of the law against such interference or attacks.

13. a. Everyone has the right to freedom of movement and residence within the borders of each state.
    b. Everyone has the right to leave any country, including his own, and to return to his country.
14. a. Everyone has the right to seek and to enjoy in other countries asylum from persecution.
    b. This right may not be invoked in the case of prosecutions genuinely arising from non-political crimes or from acts contrary to the purposes and principles of the United Nations.
15. a. Everyone has the right to a nationality.
    b. No one shall be arbitrarily deprived of his nationality nor denied the right to change his nationality.
16. a. Men and women of full age, without any limitation due to race, nationality or religion, have the right to marry and to found a family. They are entitled to equal rights as to marriage, during marriage and at its dissolution.
    b. Marriage shall be entered into only with the free and full consent of the intending spouses.
    c. The family is the natural and fundamental group unit of society and is entitled to protection by society and the State.
17. a. Everyone has the right to own property alone as well as in association with others.
    b. No one shall be arbitrarily deprived of his property.
18. Everyone has the right to freedom of thought, conscience and religion; this right includes freedom to change his religion or belief, and freedom, either alone or in community with others and in public or private, to manifest his religion or belief in teaching, practice, worship and observance.
19. Everyone has the right to freedom of opinion and expression: this right includes freedom to hold opinions without interference and to seek, receive and impart information and ideas through any media and regardless of frontiers.
20. a. Everyone has the right to freedom of peaceful assembly and association.
    b. No one may be compelled to belong to an association.
21. a. Everyone has the right to take part in the government of his country, directly or through freely chosen representatives.
    b. Everyone has the right of equal access to public service in his country.

c. The will of the people shall be the basis of the authority of government; this will shall be expressed in periodic and genuine elections which shall be by universal and equal suffrage and shall be held by secret vote or by equivalent free voting procedures.

22. Everyone, as a member of society, has the right to social security and is entitled to realization, through national effort and international co-operation and in accordance with the organization and resources of each State, of the economic, social and cultural rights indispensable for his dignity and the free development of his personality.

23. a. Everyone has the right to work, to free choice of employment, to just and favorable conditions of work and to protection against unemployment.
 b. Everyone, without any discrimination, has the right to equal pay for equal work.
 c. Everyone who works has the right to just and favorable remuneration ensuring for himself and his family an existence worthy of human dignity, and supplemented, if necessary, by other means of social protection.
 d. Everyone has the right to form and to join trade unions for the protection of his interests.

24. Everyone has the right to rest and leisure, including reasonable limitation of working hours and periodic holidays with pay.

25. a. Everyone has the right to a standard of living adequate for the health and well-being of himself and of his family, including food, clothing, housing and medical care and necessary social services, and the right to security in the event of unemployment, sickness, disability, widowhood, old age or other lack of livelihood in circumstances beyond his control.
 b. Motherhood and childhood are entitled to special care and assistance. All children, whether born in or out of wedlock, shall enjoy the same social protection.

26. a. Everyone has the right to education. Education shall be free, at least in the elementary and fundamental stages. Elementary education shall be compulsory. Technical and professional education shall be made generally available and higher education shall be equally accessible to all on the basis of merit.
 b. Education shall be directed to the full development of the human personality and to the strengthening of respect for human rights and fundamental freedoms. It shall promote understanding, tolerance and friendship among all nations, racial or religious groups, and shall

further the activities of the United Nations for the maintenance of peace.

c. Parents have a prior right to choose the kind of education that shall be given to their children.

27. a. Everyone has the right freely to participate in the cultural life of the community, to enjoy the arts and to share in scientific advancement and its benefits.

    b. Everyone has the right to the protection of the moral and material interests resulting from any scientific, literary or artistic production of which he is the author.

28. Everyone is entitled to a social and international order in which the rights and freedoms set forth in this Declaration can be fully realized.

29. a. Everyone has duties to the community in which alone the free and full development of his personality is possible.

    b. In the exercise of his rights and freedoms, everyone shall be subject only to such limitations as are determined by law solely for the purpose of securing due recognition and respect for the rights and freedoms of others and of meeting the just requirements of morality, public order and the general welfare in a democratic society.

    c. These rights and freedoms may in no case be exercised contrary to the purposes and principles of the United Nations.

30. Nothing in this Declaration may be interpreted as implying for any State, group or person any right to engage in any activity or to perform any act aimed at the destruction of any of the rights and freedoms set forth herein.

Stockholm Declaration on the Human Environment, June 1972

Principle 1 Man has the fundamental right to freedom, equality, and adequate conditions of life, in an environment of a quality that permits a life of dignity and well being, and he bears a solemn responsibility to protect and improve the environment for present and future generations.

Principle 2 The natural resources of the earth including the air, water, land, flora and fauna and especially representative samples of natural ecosystems must be safeguarded for the benefit of present and future generations through careful planning or management, as appropriate.

Principle 3 The capacity of the earth to produce vital renewable resources must be maintained and, wherever practicable, restored or improved.

Principle 4 Man has a special responsibility to safeguard and wisely manage the heritage of wildlife and its habitat that are now gravely imperiled by the combination of averse factors. Nature conservation including wildlife must therefore receive importance in planning for economic development.

Principle 5 The non-renewable resources of the earth must be employed in such a way as to guard against the danger of their future exhaustion and to ensure the benefits from such employment are shared by all mankind.

Principle 6 The discharge of toxic substances or of other substances and the release of heat, in such quantities or concentrations as to exceed the capacity of the environment to render them harmless, must be halted in order to ensure that serious or irreversible damage is not inflicted upon ecosystems ...

Principle 7 States shall take all possible steps to prevent pollution to the seas by substances that are liable to create hazards to human health, to harm living resources and marine life, to damage amenities or to interfere with other legitimate uses of the sea.

Principle 8 Economic and social development is essential for ensuring a favourable living and working environment for man and for creating conditions on Earth that are necessary for the improvement of the quality of life.

Principle 9 Environmental deficiencies generated by the conditions of underdevelopment and natural disasters pose grave problems and can best be remedied by accelerated development through the transfer of substantial quantities of financial and technological assistance as a supplement to the domestic effort of the developing countries and such timely assistance as may be required.

Principle 10 For the developing countries, stability of prices and adequate earnings for primary commodities and raw materials are essential to environmental management, since economic factors as well as ecological processes must be taken into account.

Principle 11 The environmental policies of all States should enhance and not adversely affect the present or future development potential of developing countries, nor should they hamper the attainment of better living conditions for all, and appropriate steps should be taken by States and international organizations with a view to reaching agreement on meeting the possible national and international economic consequences resulting from the application of environmental measures.

Principle 12 Resources should be made available to preserve and improve the environment, taking into account the circumstances and particular requirements of developing countries and any costs which may emanate- from their incorporating environmental safeguards into their development planning and the need for making available to them, upon their request, additional international technical and financial assistance for this purpose.

Principle 13 In order to achieve a more rational management of resources and thus to improve the environment, States should adopt an integrated and co-coordinated approach to their development planning so as to ensure that development is compatible with the need to protect and improve the human environment for the benefit of their population.

Principle 21 States have, in accordance with the Charter of the United Nations and the principles of Environmental law, the sovereign right to exploit their own resources pursuant to their own environmental policies, and the responsibility to ensure that activities within their jurisdiction or control do not cause damage to the environment of other states or of areas beyond the limits of national jurisdiction.

Principle 22 States shall cooperate to develop further the international law regarding liability and compensation for the victims of pollution and other environmental damage caused by activities within the jurisdiction or control of such states to areas beyond their jurisdiction.

Principle 24 International matters concerning the protection and improvement of the environment should be handled in a co-operative spirit by all countries, big or small, on an equal footing. Cooperation through multilateral or bilateral arrangements or other appropriate means is essential to effectively control, prevent, reduce and eliminate adverse environmental effects resulting from activities conducted in all spheres, in such a way that due account is taken of the sovereignty and interests of all States.

Declaration of Rio de Janeiro on Environment and Development, June 1992

Principle 1 Human beings are at the centre of concerns for sustainable development. They are entitled to a healthy and productive life in harmony with nature.

Principle 2 States have, in accordance with the Charter of the United Nations and the principles of international law, the sovereign right to exploit their own resources pursuant to their own environmental and developmental policies, and the responsibility to ensure that activities within their jurisdiction or control do not cause damage to the environment of other States or of areas beyond the limits of national jurisdiction.

Principle 3 The right to development must be fulfilled so as to equitably meet developmental and environmental needs of present and future generations.

Principle 4 In order to achieve sustainable development, environmental protection shall constitute an integral part of the development process and cannot be considered in isolation from it.

Principle 5 All States and all people shall cooperate in the essential task of eradicating poverty as an indispensable requirement for sustainable development, in order to decrease the disparities in standards of living and better meet the needs of the majority of the people of the world.

Principle 6 The special situation and needs of developing countries, particularly the least developed and those most environmentally vulnerable, shall be given special priority. International actions in the field of environment and development should also address the interests and needs of all countries.

Principle 7 States shall cooperate in a spirit of global partnership to conserve, protect and restore the health and integrity of the Earth's ecosystem. In view of the different contributions to global environmental degradation, States have common but differentiated responsibilities. The developed countries acknowledge the responsibility that they bear in the international pursuit of sustainable development in view of the pressures their societies place on the global environment and of the technologies and financial resources they command.

Principle 8 To achieve sustainable development and a higher quality of life for all people, States should reduce and eliminate unsustainable patterns of production and consumption and promote appropriate demographic policies.

Principle 9 States should cooperate to strengthen endogenous capacity-building for sustainable development by improving scientific understanding through exchanges of scientific and technological knowledge, and by enhancing the development, adaptation, diffusion and transfer of technologies, including new and innovative technologies.

Principle 10 Environmental issues are best handled with the participation of all concerned citizens, at the relevant level. At the national level, each individual shall have appropriate access to information concerning the environment that is held by public authorities, including information on hazardous materials and activities in their communities, and the opportunity to participate in decision-making processes. States shall facilitate and encourage public awareness and participation by making information widely available. Effective access to judicial and administrative proceedings, including redress and remedy, shall be provided.

Principle 11 States shall enact effective environmental legislation. Environmental standards, management objectives and priorities should reflect the environmental and developmental context to which they apply. Standards applied by some countries may be inappropriate and of unwarranted economic and social cost to other countries, in particular developing countries.

Principle 12 States should cooperate to promote a supportive and open international economic system that would lead to economic growth and sustainable development in all countries, to better address the problems of environmental degradation. Trade policy measures for environmental purposes should not constitute a means of arbitrary or unjustifiable discrimination or a disguised restriction on international trade. Unilateral actions to deal with environmental challenges outside the jurisdiction of the importing country should be avoided. Environmental measures addressing transboundary or global environmental problems should, as far as possible, be based on an international consensus.

Principle 13 States shall develop national law regarding liability and compensation for the victims of pollution and other environmental damage. States shall also cooperate in an expeditious and more determined manner to develop further international law regarding liability and compensation for adverse effects of environmental damage caused by activities within their jurisdiction or control to areas beyond their jurisdiction.

Principle 14 States should effectively cooperate to discourage or prevent the relocation and transfer to other States of any activities and substances that cause severe environmental degradation or are found to be harmful to human health.

Principle 15 In order to protect the environment, the precautionary approach shall be widely applied by States according to their capabilities. Where there are threats of serious or irreversible damage, lack of full scientific certainty shall not be used as a reason for postponing cost-effective measures to prevent environmental degradation.

Principle 16 National authorities should endeavour to promote the internalization of environmental costs and the use of economic instruments, taking into account the approach that the polluter should, in principle, bear the cost of pollution, with due regard to the public interest and without distorting international trade and investment.

Principle 17 Environmental impact assessments, as a national instrument, shall be undertaken for proposed activities that are likely to have a significant adverse impact on the environment and are subject to a decision of a competent national authority.

Principle 18 States shall immediately notify other States of any natural disasters or other emergencies that are likely to produce sudden harmful effects on the environment of those States. Every effort shall be made by the international community to help States so afflicted.

Principle 19 States shall provide prior and timely notification and relevant information to potentially affected States on activities that may have a significant adverse transboundary environmental effect and shall consult with those States at an early stage and in good faith.

Principle 20 Women have a vital role in environmental management and development. Their full participation is therefore essential to achieve sustainable development.

Principle 21 The creativity, ideals and courage of the youth of the world should be mobilized to forge a global partnership in order to achieve sustainable development and ensure a better future for all.

Principle 22 Indigenous people and their communities and other local communities have a vital role in environmental management and development because of their knowledge and traditional practices. States should recognize and duly support their identity, culture and interests and enable their effective participation in the achievement of sustainable development.

Principle 23 The environment and natural resources of people under oppression, domination and occupation shall be protected.

Principle 24 Warfare is inherently destructive of sustainable development. States shall therefore respect international law providing protection for the environment in times of armed conflict and cooperate in its further development, as necessary.

Principle 25 Peace, development and environmental protection are interdependent and indivisible.

Principle 26 States shall resolve all their environmental disputes peacefully and by appropriate means in accordance with the Charter of the United Nations.

Principle 27 States and people shall cooperate in good faith and in a spirit of partnership in the fulfillment of the principles embodied in this Declaration and in the further development of international law in the field of sustainable development.

Specialized Summit of the Americas on Sustainable Development, December 1996

1. We, the elected Heads of State and Government of the Americas, gathered in Santa Cruz de la Sierra as decided at the Summit of the Americas held in Miami in 1994, reaffirm our determination to move forward toward sustainable development and to implement the decisions and commitments set forth in the Rio Declaration and Agenda 21, which were adopted at the United Nations Conference on Environment and Development, held in Rio de Janeiro in 1992.

   We also reaffirm the commitments undertaken in the Declaration of Principles and the Plan of Action of the Summit of the Americas.

   We undertake to promote the agreements reached at the Global Conference on the Sustainable Development of Small Island Developing States, held in Barbados in 1994, and recognize the importance of the principles enunciated at recent United Nations conferences concerning sustainable development.

   We support the efforts launched at the hemispheric, regional, and subregional levels, such as the Central American Alliance for Sustainable Development, the North American Agreement on Environmental Cooperation, the Treaty for Amazonian Cooperation, and the Permanent South Pacific Commission.

2. We reaffirm that human beings are entitled to a healthy and productive life in harmony with nature and, as such, are the focus of sustainable development concerns. Development strategies need to include sustainability as an essential requirement for the balanced, interdependent, and integral attainment of economic, social, and environmental goals.

3. One essential feature of the Americas is their natural and cultural diversity. Our countries share a unique and rich political tradition grounded in democratic values and significant potential for economic growth and technological development in a context of open, market-based economies. These characteristics are of fundamental importance for the promotion of economic development and social welfare and for the preservation of a healthy environment.

We will adopt policies and strategies that will encourage changes in production and consumption patterns in order to attain sustainable development and a better quality of life, as well as to preserve our natural environment and contribute to the alleviation of poverty.

We reaffirm our commitment to the fundamental principle of the Charter of the Organization of American States, restated at the Summit of the Americas, that representative democracy is essential for peace, justice, and development. Sustainable development requires that we strengthen and promote our democratic institutions and values.

4. Recognizing that globalization, efforts toward integration, and the complexity of environmental issues pose challenges and offer opportunities to the countries in the Hemisphere, we pledge to work together.

5. We recognize that the needs and responsibilities facing the countries of the Hemisphere today are diverse. Sustainable development does not assume that all the countries are at the same level of development, have the same capabilities, or can necessarily use the same model to attain it. In view of the different contributions to global environmental degradation, states have common but differentiated responsibilities in the global quest for sustainable development. We should make efforts to ensure that the benefits of sustainable development reach all countries in the Hemisphere, in particular those that are less developed, and all segments of our populations.

   We will give special attention to the small island states, whose environmental vulnerability, especially with regard to natural disasters, is greater owing to their geographic situation, their size, and the scale of their economies, among other factors.

6. The alleviation of poverty is an integral part of sustainable development. The benefits of prosperity will only be attained through policies that address the interrelationship between human beings and nature. In developing policies and programs for sustainable development, special attention should be given to the needs of indigenous people, minority communities, women, youth, and children and to facilitating their full participation in the development

process. The living conditions of persons with disabilities and the elderly also merit special attention.

7. We will establish or strengthen our programs, policies, and institutional frameworks in support of sustainable development objectives. National efforts should be complemented by ongoing international cooperation in furtherance of the commitments made at the Rio conference related to financial resources, and the transfer of technology on fair and favorable terms, including preferential terms, as mutually agreed.

8. We will support and encourage, as a basic requisite for sustainable development, broad participation by civil society in the decision-making process, including policies and programs and their design, implementation, and evaluation. To this end, we will promote the enhancement of institutional mechanisms for public participation.

9. This Summit Conference on Sustainable Development is the cornerstone of a partnership for cooperation among the states of the Americas in their common pursuit of a higher quality of life for their peoples, founded on integrated and complementary economic, social, and environmental objectives.
Taking the current experience of our countries and region as a point of departure, we hereby frame a plan of action that will commit the states to timely action and ensure the availability of the resources needed for that purpose.

10. In keeping with the principles stated above, we emphasize the following points regarding application of the Plan of Action for the Sustainable Development of the Americas:

a. Equitable economic growth
Implement effective and ongoing measures to ensure that the international economic and financial system supports the growth of local economies and their sustainable development with a view to establishing greater social justice for all of our peoples.

Reinforce the mutually supportive relationship between trade and the environment by acting to conserve the environment, while safeguarding an open, equitable, and nondiscriminatory multilateral trade system, taking into account the efforts currently being deployed in this field by the Committee on Trade and Environment of the World Trade Organization. We recognize the great need of our countries to improve access to markets while maintaining effective and appropriate environmental policies. In this regard, we will avoid hidden trade restrictions, in accordance with the General Agreement on Tariffs and Trade/World Trade Organization (GATT/WTO) and other international obligations.

Full participation by the private sector--especially small, medium-sized, and micro-enterprises, as well as cooperatives and other forms of productive organization--in a sustainable development strategy essential to take advantage of its resources and dynamism. This strategy should balance comprehensive policies to address environmental and development problems.

b. Social dimensions

There is an urgent need to intensify efforts to reduce the poverty and the marginalization which broadly affect our societies, and especially women and children. We will promote, through the relevant measures and programs, including those established in the Plan of Action, adequate levels of nutrition, a greater degree of food security, equitable and effective access to basic health care and drinking water and to employment and housing, and we will seek to ensure pollution control and a clean environment for all people, taking into account, in particular, the most vulnerable groups.

We will also develop strategies that value human dignity while respecting and fostering the cultural diversity of our societies, gender equity, and educational programs promoting peace, democracy, and respect for nature, with special attention to children and young people.

In this context, the principles and priorities established in the Pan American Charter: Health and Environment in Sustainable Human Development will be put into practice as appropriate.

c. A healthy environment

Planning and decision-making for sustainable development require understanding and integrating environmental considerations, as well as social and economic factors. We will assess the environmental impact of our policies, strategies, programs, and projects nationally and in the framework of international agreements to ensure that adverse environmental effects are identified, prevented, minimized, or mitigated, as appropriate.

d. Public participation
We will promote increased opportunities for the expression of ideas and the exchange of information and traditional knowledge on sustainable development between groups, organizations, businesses, and individuals, including indigenous people, as well as for their effective participation in the formulation, adoption, and execution of decisions that affect their lives.

e. The development and transfer of technology
The development, adoption, adaptation, and application of environmentally sound, effective technology play an important role in ensuring sustainable development.

To this end, efforts to promote the transfer of, and access to, appropriate technology should continue in the Hemisphere. We recognize the important role played by market-based mechanisms and will promote opportunities for technology transfer through training and cooperative work programs and through improved access to sources of information. In addition, we will strengthen national scientific and technological capacities, complemented by international cooperation.

f. Financing
Implementation of the initiatives set forth in the Plan of Action requires the mobilization of financial resources in keeping with the commitments made at the Rio Summit. These should be complemented with innovative financing mechanisms.

In this context, we highlight the importance of international organizations and financial institutions in strongly supporting the efforts of the Hemisphere.

g. Strengthening of the legal framework
Relations between countries of the Hemisphere, within the

framework of this partnership for sustainable development, will be grounded in the rules and principles of international law. We will consider the progress in international environmental law and promote the reform and modernization of national laws, as appropriate, to reflect sustainable development concepts. We will also develop national mechanisms for effective enforcement of applicable international and national laws and provisions.
We will seek to secure ratification of, or accession to, international instruments on sustainable development and will fulfill all commitments made therein.

Thus we sign the Declaration of Santa Cruz and adopt the Plan of Action for the Sustainable Development of the Americas on this seventh day of December in the year one thousand nine hundred and ninety-six, in Spanish, French, English, and Portuguese.

Millennium Declaration Values September 2000

Freedom	Men and women have the right to live their lives and raise their children in dignity, free from hunger and from the fear of violence, oppression or injustice. Democratic and participatory governance based on the will of the people best assures these rights.
Equality	No individual and no nation must be denied the opportunity to benefit from development. The equal rights and opportunities of women and men must be assured.
Solidarity	Global challenges must be managed in a way that distributes the costs and burdens fairly in accordance with basic principles of equity and social justice. Those who suffer or who benefit least deserve help from those who benefit most.
Tolerance	Human beings must respect one other, in all their diversity of belief, culture and language. Differences within and between societies should be neither feared nor repressed, but cherished as a precious asset of humanity. A culture of peace and dialogue among all civilizations should be actively promoted.
Respect for nature	Prudence must be shown in the management of all living species and natural resources, in accordance with the precepts of sustainable development. Only in this way can the immeasurable riches provided to us by nature be preserved and passed on to our descendants. The current unsustainable patterns of production and consumption must be changed in the interest of our future welfare and that of our descendants.
Shared responsibility	Responsibility for managing worldwide economic and social development, as well as threats to international peace and security, must be shared among the nations of the world and should be exercised multilaterally. As the most universal and most representative organization in the world, the United Nations must play the central role.

Millennium Development Goals (MDGs) September 2000

By the year 2015, all 191 UN Member State have pledged to:

Goal 1 Eradicate extreme poverty and hunger
- Reduce by half the proportion of people living on less than one U.S. dollar per day.
- Reduce by half the proportion of people who suffer from hunger.

Goal 2 Achieve universal primary education
- Ensure that all boys and girls complete a full course of primary schooling.

Goal 3 Promote gender equality and empower women
- Eliminate gender disparity in primary and secondary education preferably by 2005, and at all levels by 2015.

Goal 4 Reduce child mortality
- Reduce by two-thirds the mortality rate among children under five.

Goal 5 Improve maternal health
- Reduce by three-quarters the maternal morality ratio.

Goal 6 Combat HIV/AIDS, malaria and other diseases
- Halt and begin to reverse the spread of HIV/AIDS.
- Halt and begin to reverse the incidence of malaria and other major diseases.

Goal 7 Ensure environmental sustainability
- Integrate the principles of sustainable development into country policies and programmes; reverse loss of environmental resources.
- Reduce by half the proportion of people without sustainable access to safe drinking water.
- Achieve significant improvement in lives of at least 100 million slum dwellers, by 2020.

Goal 8 Develop a global partnership for development

- Develop further an open trading and financial system that is rule-based, predictable and non-discriminatory. Includes a commitment to good governance, development and poverty reduction – nationally and internationally.
- Address the least developed countries' special needs. This includes tariff – and quota-free access for their exports; enhanced debt relief for heavily indebted poor countries; cancellation of official bilateral debt; and more generous official development assistance for countries committed to poverty reduction.
- Address the special needs of landlocked and Small Island developing States.
- Deal comprehensively with developing countries' debt problems through national and international measures to make debt sustainable in the long-term.
- In cooperation with the developing countries, develop decent and productive work for youth.
- In cooperation with pharmaceutical companies, provide access to affordable essential drugs in developing countries.
- In cooperation with the private sector, make available the benefits of new technologies – especially information and communications technologies.

Appendix 2

Selected Intergovernmental Environmental Agreements

Early Agreements

1. Charter of the United Nations (1945)
2. Declaration of the United Nations Conference on the Human Environment (1942)
3. The Universal Declaration of Human Rights (1944)
4. The Convention on the Rights of the Child
5. Convention on the Elimination of all Forms of Discrimination Against Women
6. UNESCO Convention Concerning the Protection of the World Cultural and Natural Heritage

UNCED (1992) Preparations

7. A/Res/42/186 of 11 December 1987
8. A/Res/42/187 of 11 December 1987
9. A/Res/44/228-85 of 22 December 1989

UNCED Related Documents

10. The Rio Declaration on Environment and Development
11. *Agenda 21*: Global Programme of Action on Sustainable Development
12. Statement of Principles for the Sustainable Management of Forests
13. Declaration of Barbados and the Programme of action for the Sustainable Development of Small Island Developing States
14. Programme of Action adopted at the Special Session of the General Assembly to Review the Implementation of *Agenda 21* (Earth Summit+5)
15. United Nations Millennium Declaration and Goals

Social Issues

16. WHO-policy Health for All in the 21st Century
17. The Habitat Agenda and the Istanbul Declaration on Human Settlements
18. Programme of Action of the International Conference on Population and Development (1994)
19. The Report of the World Summit for Social Development (Copenhagen, 1995)

Economic Issues

20. Final Act of the Uruguay Round and Related Agreements
21. The Marrakech Ministerial Decision on Trade and Environment
22. WTO document on the relationship between the provision of the multilateral trading system and measures for environmental purposes, including those pursuant to multilateral environmental agreements
23. Further WTO documents on trade and environment

Natural Resources

24. Rome Declaration on World Food Security and the World Food Summit Plan of Action
25. Report of the World Food Summit (1996)
26. World Declaration on Nutrition/Plan of Action for Nutrition
27. UN Framework Convention on Climate Change
28. The Montreal Protocol
29. Vienna Convention for the Protection of the Ozone Layer (1985)
30. Convention on Biological Diversity
31. Convention on International Trade in Endangered Species of Wild Fauna and Flora
32. Convention on the Conservation of Migratory Species of Wild Animals and further agreements under this convention
33. Convention on Wetlands of International Importance Especially as Waterfowl Habitat
34. United Nations Convention to Combat Desertification in Countries Experiencing Serious Drought and/or Desertification, Particularly in Africa

Hazardous Wastes

35. Basel Convention on the Control of Transboundary Movements of Hazardous Wastes and their Disposal
36. Bamako Convention on the Ban of the Import into Africa and the Control of Transboundary Movement and Management of Hazardous Wastes within Africa
37. Joint Convention on the Safety of spent Fuel Management and on the Safety of radioactive Waste Management
38. Convention on Supplementary Compensation for Nuclear Damage
39. Convention on Early Notification of a Nuclear Accident
40. Convention on Physical Protection of Nuclear Material
41. Convention on Nuclear Safety
42. Vienna Convention on Civil Liability for Nuclear Damage

Oceans

43. UN Convention on the Law of the Sea
44. IMO Conventions in the field of marine safety, prevention of marine pollution
45. FAO Code of Conduct for Responsible Fisheries
46. Kyoto Declaration and Plan of Action
47. Regional Seas Action Plans

Appendix 3

Major Civil Society Alternative Agreements

(Texts of these documents may be found at
http://www.igc.org/habitat/treaties/index.html)

Declarations and General Principles

1. People's Earth Declaration
2. Rio de Janeiro Declaration
3. The Earth Charter
4. Ethical Commitments to Global Ecological Posture and Behavior

Education, Communication and Cooperation

5. Treaty on Environmental Education for Sustainable Societies and Global Responsibility
6. Communication, Information, Media and Networking Treaty
7. Treaty for Non-Governmental Organization Cooperation and Sharing of Resources
8. Treaty on a Technology Bank Solidarity System for Technological Exchange
9. Rio Framework Treaty on NGO Global Decision Making
10. Code of Conduct for NGOs

Alternative Economic Issues

11. Treaty on Alternative Economic Models
12. Alternative Treaty on Trade and Sustainable Development
13. Debt Treaty
14. Treaty of the People of the Americas
15. Capital Flight and Corruption Treaty
16. Treaty on Transnational Corporations: Democratic Regulation of their Conduct

Consumption, Poverty, Food and Subsistence

17. Treaty on Consumption and Lifestyle
18. Poverty Treaty
19. Food Security Treaty
20. Sustainable Agriculture Treaty
21. Fresh Water Treaty
22. Fisheries Treaty

Climate, Energy and Waste

23. Alternative Non-Governmental Agreement on Climate Change
24. Treaty on Energy
25. Treaty on Waste
26. Treaty on the Nuclear Problem

Land and Natural Resources

27. Forest Treaty
28. Treaty Regarding Arid and Semi-Arid Zones
29. Treaty on "Cerrados" (Scrubland)

Marine and Ocean Issues

30. Pollution of the Marine Environment
31. Minimizing Physical Alteration of Marine Ecosystems
32. Protecting the Sea from Global Atmospheric Changes
33. Marine Protected Areas
34. Resolution Concerning Guanabara Bay: Humankind's Heritage

Biodiversity and Biotechnology

35. Citizens' Commitments on Biodiversity
36. Marine Biodiversity Treaty
37. Draft Protocol on Scientific Research Components for the Conservation of Biodiversity
38. Citizen's Commitments on Biotechnology

Cross-Sectoral Issues

39. A Global Women's Treaty for NGOs Seeking a Just and Healthy Planet
40. Treaty on Population, Environment and Development
41. Youth Treaty
42. Treaty in Defense and Protection of Children and Adolescents
43. International Treaty between Non-Governmental Organizations and Indigenous Peoples
44. Treaty Against Racism
45. Treaty on Militarism, the Environment and Development
46. Treaty on Urbanization

Appendix 4

Documenting Governance

Although the purpose of any GEG meeting is to produce a set of decisions to guide activities designed to meet the stated objectives of that meeting, it sometimes appears that a meeting's purpose is little more than to produce a set of documents of ever increasing number. To keep track of all these documents, from three to five lines of document symbols or codes – a combination of letters and numbers – are placed in the upper right corner of the first page that indicate just what the document is: the responsible body and/or subsidiary body, the meeting, identification number, where it is in the process, when it was written or approved, and in what languages. There are similarities among the many agencies and agreements but there are also differences – for example, the UN uses symbols that are composed of upper case letters while other agencies may use lower case and the symbols used by the Organization of American States are in Spanish even though the document may be in English, French or Portuguese. How to decipher just what all the abbreviations mean is one of the first hurdles a beginner has to make – both to keep track of what is going on in a meeting and to find the document you are looking for in the future. We will present a few of the more common abbreviations here. Others can be found in the "Condensed Glossary of Selected Terms, Concepts, Jargon, Acronyms and Abbreviations Used in Global Environmental Governance" section of this book. The United Nations explains its documentation symbols at http://www.un.org/depts./dh/resguides/symbol.htm.

Generally, units of the code are divided by a diagonal line [/] (and sometimes by a period [.] or hyphen [-]). The first set of letters identify the parent organ or body and the second set identifies the subsidiary body responsible for the document. The next set identifies the meeting or activity and the number immediately following identifies the session of the meeting where the document is being, or was, presented and whether it is the original, an addendum, amendment, or correction. Other symbols indicate the nature of the document's distribution (restricted, general, internal, limited,) and additional information such as whether it is for information only, whether it is a statement by a named member party or observer, a verbatim or summary record, a resolution, a working paper or a petition. Information may also be found elsewhere on the cover sheet that indicates the full name of the sponsoring institutions, location and date of the meeting, and the title of the paper. The following is an example of what would appear in the upper right corner of the first page (the main line of which is often repeated on each page of the entire document):

Dist
GENERAL
UNEP/CBD/COP/3/Inf.20
20 September 1996
ORIGINAL: ENGLISH

This notation says that the document was originally written in English for general distribution on September 20, 1996 and is the twentieth informational paper in a series prepared for the third meeting of the Conference of the Parties to the Convention on Biological Diversity being sponsored by the United Nations Environment Programme. This is an easy example. When they get a lot more difficult, and they do, don't panic. You can go through the following table and maybe figure it out. Or you can just ask your neighbor what they mean. Of course, your neighbor won't know either, but it is always more comforting when you realize that you are not alone.

Document Identification Symbols

Abbreviation	**Full Name/Description**	**Organization(s)**
A	Assembly, General Assembly	GEF, UN
Add.	Addendum, the second part of a document previously submitted	UN
AC	Ad hoc committee	UN
AG	General Assembly (*Asamblea General*)	OAS
Amend.	Alteration, by decision of a competent authority, of a portion of an adopted text	UN
C	Council	GEF
C	Standing/permanent/main committee	UN
CN	1. Conference Notes, 2. Committee Notes, 3) Commission	UN
Conf.	Conference paper	
CONF.	Conference paper	UN
Corr.	Corrigendum, corrections to a document	
CP	Pre-Conference Documents (Provisional and regular documents and agenda)	UN
CP	Permanent Council (*Consejo Permanente*)	OAS

CRP	Conference Room Papers (Working documents for use during negotiations)	
Dist.	Distribution of the document	
DP	United Nations Development Programme	UN
E	Economic and Social Council	UN
E	Extraordinary session	OAS
E	1. English, 2. Edited, 3. Executive	UN
ED	Executive Directive	UNEP
G	General Distribution	UN
GC	Governing Council	UN
GET/PMA	Special Working Group on Environmental Protection	OAS
GT	Working Group (*Grupo de Trabajo*) of the Permanent Council	OAS
IDP	Internal Discussion Paper	WB
IDR	In-depth reviews (of national communications)	WB
Inf.	Information document	
INF	Information series	UN
INF. docs	Information documents (Background)	
JD	Board of Directors (*Junta Directiva*)	OAS
LC	Latin America and the Caribbean	
L	Limited distribution	
L. docs	Limited documents (Draft reports and texts)	
Misc. Docs	Miscellaneous documents (Views of Parties/ Observers; participant lists)	
NC	National Communication	UN
Non-paper	Unofficial document (Informal, in-session documents to assist negotiations)	UN
OEA	OAS	OAS
OP	Operational procedure	WB
PC	Preparatory committee	UN
PET	Petitions	UN
PRST	Statements by the President of the Security Council	UN
PV	Verbatim records of meetings	UN
R	Restricted distribution; restricted access	UN
Rec.	Recommendation (Rec.1, Rec.2, etc)	
RES (or Res)	Resolution	
Rev.	Revision (Revised document)	
S	Security Council	UN

SA	Synthesis of the minutes	OAS
SC	Subcommittee	UN
Ser.	Series	
Ser. A	Multilateral agreements, conventions and treaties open for signature in the OAS	OAS
Ser. B	Agreements to which the OAS is party	OAS
Ser. G	Permanent Council	OAS
Ser. L	Inter-American Specialized Organizations	OAS
Ser. M	Bilateral and regional conventions deposited at the OAS	OAS
Ser. P	General Assembly	OAS
SR	Summary records of meetings	UN
SS	Special Session	
ST	Secretariat	UN
Sub.	Subcommission	UN
Summary	Summarized version	
TD	United Nations Conference on Trade and Development	UN
TP	Technical papers	
UNEP	United Nations Environment Programme	UN
WG	Working group	
WP	Working papers	UN

Appendix 5

Random Definitions

The following are definitions given by graduate students and professionals working in the "environmental" offices of government and some NGOs when asked to give their definitions to the terms "environment," "ecology," "ecosystem," and "sustainable development." The questions were asked at the beginning of short courses in spatial planning, environmental governance, and natural resource management. (Source: Personal files of Richard E. Saunier)

Environment

- National parks, wilderness, and wildlife.
- A combination of many different elements occurring both naturally and artificially.
- Surroundings, relating to land, water, and air systems.
- The physical and chemical domains in which living and nonliving things exist.
- The bios surrounding a particular area.
- All things, be it natural or man made, that surround a person(s) or are within a defined space.
- Plants, animals, water, soils, air, ...everything involved in our surrounding area. All inter-reliant. Therefore in any one thing affect others.
- Group of ecosystems.
- The flora, fauna, geological, hydrological and atmospheric components functioning together.
- The external factors that affect the going concern of any system.
- The group of natural elements such as climatology, soil, water, plants, and animals.
- Ensemble of place where people, animals and all living things interact with each other.
- Many aspects occur in small or large area like, pollution etc.
- The surroundings that living things live in. This includes inorganic and organic matter.
- Any surrounding habitable or non-habitable by living things.
- The amount of factors such as climate, vegetation, soils, etc. that are around.

- In a broad sense, the concept of environment refers to the surroundings that have some influence on the life of the subjects in it.
- The media that contains many different parts working together or depending on each other in a specific area.
- Man's (or any living thing) surroundings.
- ...the area or space in which human beings interact.
- Ones' environment is that which surrounds him (her).
- A combination of natural inputs including animals (both human and wildlife), water, air, land etc. without environment there is no life.
- The care of the surroundings. Protecting the country resources.
- Every aspect of surroundings that in some way affects all forms of life – human, plant, animal, etc.
- The natural and artificial surroundings.
- The environment is everything.
- The assemblage of many ecosystems which produce the 5 renewable natural resources.
- Space.
- Atmosphere. Everything to do with nature – land, sun, air, sea. Man and his surroundings.
- Collection of natural conditions that condition human activities.
- Space that surrounds us.
- Main characteristics that are present in a given place.
- Group of elements that form the natural surroundings that are interdependent in terms of their development. Ecosystem.
- The physical-social-economic surroundings including the interrelationships that are established between them, in which life is developed and which influences a being's life in general.
- Natural environment includes water, air, land etc. animals and people also. (2) Broader definition to include homes, education, and other human development indicators.
- All of nature that surrounds and affects the entire universe.
- What I want or need, when I want or need it, where I want or need it.
- The interdependent whole made up of divers entities.
- The reason to be alive.
- Where we live, work, and play.
- The space in which we live and that sustains us.
- All physical and social aspects influencing and sustaining life.

Ecology

- The science that study the different associations of organisms and their relationship with the environment.
- Science which goal is to study the environment.
- It means there are many aspects of ecology in the sea, river, lakes, etc.
- A combination of ecosystems and how they interact with one another both in the ecosystems and between the ecosystems.
- The study of living things and their relationship with the environment.
- The science that study the environment.
- The discipline which is related to the weather, in other words to the way of live of living beings in a given environment.
- The science of dealing with living things.
- The study of environment and all its inhabitants.
- The study of the interrelationship of living things with their environment.
- The scientific discipline [or] study of earth's different systems and environments.
- The study of the earth's natural resources and their interactions.
- The study of the interactions between the biotic and abiotic constituents of an environment.
- The study of the environment surrounding and the way it works.
- The study of the environment.
- The study of ecosystems (plants, animals within the ecosystem).
- Science of particulars (flora/fauna).
- Science of understanding the interactions of plants, animals, environment.
- The science that take care of matters regardless of natural resources.

Ecosystem

- A classification for an area including plants, animals, trees and environment. They are areas that have specific characteristics in terms of ecological significance.
- The natural environment of the earth and the interaction with animals and other living things.
- The biotic and abiotic constituents of an environment.
- The components that make up a natural environmental system.
- The natural environment within a particular area.
- A natural system and all related plant/animal species within the system.

Also the environment of the system (i.e. weather/climate, geography).

- Balanced concentration of particulars.
- The plants, animals, water, land, etc. that comprise a definable ecological system. For example, the everglades ecosystem is comprised of water flowing from N to S, alligators, gar, anhingas, coral/limestone based soils, subtropical climate, etc.
- The group of natural factors that are components in an occurrence, that can be measured.

Sustainable Development

- The concept of having development be more geared toward environmental, public, social issues to foster better "smart growth" in the future, trying to keep or make better what is being built in the future.
- Utilization of the earth's resources in a manner that allows for controlled depletion. Using these resources in a way which allows the environment to recover as quickly as possible, thereby allowing for further development.
- May be better entitled, "sustainable existence," because to "develop" something in human terms is inherently to exploit and drain the natural resources provided.
- The idea to perpetuate the existing ecosystem and environment at its ideal, optimal level for our existence to survive.
- Development that is self-reliant and is able to continue without outside intervention.
- A level of development which does not deteriorate other things (i.e. sustain current habitat, sustain current runoff quantities) levels of development beyond this negatively impact surrounding area.
- Balanced environments.
- Development that preserves biological diversity (doesn't cause extinctions/preserves unique ecosystems and preserves enough "green space" for current and future generations.
- Continued improvement of any system.
- Improving the quality of life.
- Improving the human condition.

Authors Biographies

RICHARD E. SAUNIER, Ph.D. was the Senior Environmental Management Advisor in the Department of Regional Development and Environment of the Organization of American States from March 1975 until his retirement in December 1996. From 1970 to 1975, he held staff positions with the U.S. Peace Corps in Paraguay, Peru and Washington, D.C. and, during 1967-69, he was visiting professor of forestry at the Universidad Austral de Chile in Valdivia, Chile. He holds degrees from Colorado State University (B.S.) in forestry and range management, and from the University of Arizona (M.S., Ph.D.) in range ecology and watershed management. His interests and recent experience are in environmental management policy and program evaluation and in the identification and mediation of resource use conflicts. Together with Richard A. Meganck, he is the co-editor, of *The Bottoms Up of International Development*, published by Infinity Publishing in 2002, and *Conservation of Biodiversity and the New Regional Planning* published by the OAS and the IUCN in 1995.
<Rsaunier@comcast.net>

RICHARD A. MEGANCK, Ph.D. is the Rector of the UNESCO-IHE Institute for Water Education in Delft, The Netherlands. Prior to his appointment in The Netherlands, Dr. Meganck was the Director of Sustainable Development and Environment at the Organization of American States. His career spans thirty years in international development and the management of natural resources in both the United Nations and Inter-American systems during which time he has undertaken more than three hundred technical and administrative missions to ninety-five countries throughout the world. He holds a B.S. and M.S. degrees from Michigan State University in natural resource development policy and watershed management and a Ph.D. from Oregon State University in natural resource management. Dr. Meganck began his professional career as a faculty member in the College of Forestry at Oregon State University and served as a Peace Corps volunteer in Colombia in 1971-1972.
<R.Meganck@unesco-ihe.org>

Kenton R. Miller Biography

DR. KENTON R. MILLER has held the highest positions in international conservation. Beginning in 1976, he served 7 years as Chairman of IUCN's World Commission on Protected Areas. From there he was elected to be Director General of IUCN where he served 5 years. In 1982 Dr. Miller was Secretary General of the 3d World National Parks Congress in Bali, Indonesia. As WRI's Director of the Biological Resources Program, he lead the cooperative project between UNEP, IUCN, and WRI, which developed the Global Biodiversity Strategy, launched at the 4th World Parks Congress in Caracas, Venezuela, in 1992. Following his Ph.D. studies at the State University of New York, College of Forestry and Environmental Sciences, he began his career as a field researcher and post graduate educator in Central and South America under the auspices of FAO which provided him with the opportunity to practice wildland management and national park planning throughout the region. Subsequently, his conservation activities carried him to all continents including Antarctica. Later, as Associate Professor at the School of Natural Resources, University of Michigan, he established the Center for Strategic Wildland Studies. Presently, Dr. Miller is Vice President for International Conservation and Development, at the WRI. In 2001, he was again elected to chair IUCN's WCPA for a four-year term. He is recipient of the Golden Ark award, and an Honorary Doctorate from the International University in Costa Rica.
<Kenton@wri.org>

Joyeeta Gupta Biography

DR. JOYEETA GUPTA has Bachelors' degrees in economics (Delhi University) and law (Gujarat University) India, a Master's degree in law from Harvard Law School, USA and a Ph.D. in law from the Vrije Universiteit in Amsterdam, the Netherlands. Since 1993, she has been working at the Institute for Environmental Studies at the Faculty of Earth and Life Sciences Vrije Universiteit. She holds a part time position as Professor in Policy and Law of Water Resources and the Environment at the UNESCO-IHE Institute for Water Education in Delft. Dr. Gupta's areas of expertise are in the multidisciplinary (legal, political and policy) analysis of international environmental and water agreements and multi-level environmental governance. In particular, her focus has been on the climate change regime, which she has examined from the North-South perspective and the perspective of Europe. She has also looked at domestic implementation issues especially in relation to energy in Brazil, China, India, Indonesia and Kenya. Although her area of research thus far has been climate change, she has been teaching water law and politics for the last twelve years. Joyeeta Gupta has published extensively in the area of environmental governance, climate change and North-South relations. She is Editor-in-Chief of *International Environmental Agreements: Politics, Law and Economics*, is on the editorial board of several important journals, and serves on a number of international scientific panels and boards.
<J.Gupta@unesco-ihe.org>

"To the reader of C.H.A.O.S.S.: Given the fluid nature of the GEG process, a book like this, particularly one containing a glossary and acronym listing must always be considered a work in progress. Next week or next year negotiations will begin on yet another convention or agreement, logically generating its own definitions and acronyms. We intend to publish an updated version of this text in the future. Therefore we ask you to send us you suggestions for inclusions and please provide a complete source for this information. Rsaunier@comcast.net; R.meganck@unesco-ihe.org"